Critique and Totality

SUNY Series in Contemporary Continental Philosophy
Dennis J. Schmidt, Editor

Critique and Totality

Pierre Kerszberg

State University of New York Press

Published by
State University of New York Press, Albany

Printed in the United States of America

For information, address State University of New York
Press, State University Plaza, Albany, N.Y., 12246

Production by Diane Ganeles
Marketing by Fran Keneston

Library of Congress Cataloging-in-Publication Data

Kerszberg, Pierre.
Critique and totality / Pierre Kerszberg.
p. cm.—(SUNY series in contemporary continental philosophy)
Includes bibliographical references and index.
ISBN 0-7914-3189-4 (hardcover: alk. paper).—ISBN 0-7914-3190-8 (pbk. : alk. paper)
1. Kant, Immanuel, 1724-1804—Contributions in cosmology.
2. Cosmology—History—18th century. 3. Cosmology. I. Title.
II. Series.
B2799.C8K47 1997
113′.09—dc20 96-3312
CIP

10 9 8 7 6 5 4 3 2 1

Contents

CHAPTER 1

Totality, Finitude, and Division

Kant and the Cosmic Concept of Philosophy

At the crucial moment when Kant, in the *Critique of Pure Reason*, characterizes the ultimate originality of critical philosophy, he seems to retreat before the extraordinary implications of his discovery. Indeed, what he asserts to be the major philosophical breakthrough of modern times consists in substituting for the scholastic concept of philosophy an understanding of philosophy as a *cosmic concept* (*conceptus cosmicus*, or *Weltbegriff*).[1] The overarching significance of the cosmic concept cannot be underestimated, since in the *Logic* Kant goes on to identify its sense with the *cosmopolitical* (*weltbürgerlichen*) concept, an expression which he uses to refer to two ultimate aspects of human existence, insofar as it is torn between inwardness and outwardness: between feeling which goes beyond discourse, and the system of states which will organize their peaceful coexistence.[2] The substitution of the cosmic concept for the scholastic, however, did not coincide with an advance, but rather the advance is itself a return, or recurrence to an earlier view. This view is rooted in the true and most dignified sense of philosophy that prevailed before the Scholastics emerged. The crucial significance of the cosmic concept for any possible philosophy lies in this double movement: a withdrawal which is supposed to reflect what Kant took to be (as he puts it poignantly at the end of the *Critique of Pure Reason*) the last path still open to us.

The scholastic tradition, according to Kant, sought no more than to perfect the logical basis of knowledge accumulated by past mathematicians, past physicists, or past logicians. On the other hand, the ideal philosopher, such as the wise man of the Stoics (A569/B597), is ideal only because he prescribes a perfection that can never have been attained in practical life. As it turns out, his philosophy "is the science of the relation of all knowledge to the essential ends of human reason" (A839/B867), which bestows upon it a dignity or absolute value that is undiminishable. By contrast, scientists may well be less ambitious; but they do not really know what they are doing because the inner nature of their object, the specificity of their method, and the very possibility of their science remain unknown to them (A725/B753). Scientists make use of a concept without becoming aware of its origin. And so the scholastic concept of philosophy finds at least a modicum of justification in that it throws light, however dimly, on this origin, thereby saving the scientist the embarrassment (within acceptable limits) of remaining in the uncomfortable position of a mere "artisan" or "craftsman" of reason. (The German word is *Künstler,* which cannot be rendered here as artist, given the contemporary meaning of the term.) But this remains a doctrine of skill, whereas the philosopher who lives in accordance with the cosmic concept does not even have the feeling of being an artisan. He is a lawgiver of human reason, inasmuch as he looks to ends which cannot be enclosed within finished products, since they can only be expressed as the unconditioned which human reason aims at.

Kant then goes on to warn against yet another scholastic distortion, which has been responsible for casting the ideal philosopher only as a moralist (A840/B868), thereby unduly restricting the scope of philosophy's search for ends to moral life and practical freedom. In fact, the philosopher as lawgiver of human reason rules over *two distinct* objects: nature and freedom. But just how are we to understand Kant's insistence on the duality of philosophy's ultimate interests?

In his commentary on the Kantian elaboration of critical philosophy as a cosmic concept, Heidegger has proposed that Kant could not see its ultimate significance.[3] Scholastic philosophy is scientific philosophy, which is not preoccupied by the worldview which animates it as the supposed goal and nature of philosophy itself. A worldview is not a matter of theoretical knowledge, nor even a part of it in some special sense. Rather, it pertains to the bedrock of meaning which guides the current affairs of human life, and which are most of the time a mixture of superstition and knowledge, sober reason and dereliction. In its true sense, the clarification of a worldview has as its task the circumscription of the characteristics which belong to the essential nature of the human Dasein. But the popular conception of life in modern times is gravely misled, because it is fraught with a double perversion. In order to provide us with a minimum of guide mark, it makes the demand on philosophy that the worldview should itself be constructed scientifically. This demand, in turn, blocks the way to the more authentic task of philosophy, which deals with what every positing of beings, even the positing done from the perspective of a worldview, must already presuppose ontologically. Whether scientific or not, a worldview philosophy is, strictly speaking, philosophically impossible. To be sure, Heidegger goes on to argue, Kant's distinction between the two senses of philosophy has the merit of showing that the scientific construction of a worldview cannot lead anywhere, because it perverts philosophy's central concern with questions as to the ends and limits of human existence. Kant, however, could not make the second and decisive step of developing these questions in accordance with their proper sense. We can stay with Kant's explication of existence for a time, but eventually we will have to part ways with him, because his demarcation of the concept of existence (through perception or positing) remains itself unclarified.

In Heidegger's interpretation of Kant, as presented in the famous Kant book of 1929, the whole critical project must be credited with the most profound level of questioning. Kant does not simply examine the limits of human knowledge, but interrogates the very possibility of metaphysics. Whether Kant himself achieves full clarification of this task is a subordinate question; what is important is that Kant recognized its necessity.[4] In the *Critique*, Kant does not simply put in question the *metaphysica specialis*, that is, the possibility of going back to the unconditioned (the three fundamental beings: God, the soul, and the world), but more profoundly the *metaphysica generalis*, that is, the possibility of an ontological fore-understanding which accounts for our encountering being as such. Kant's critical philosophy raises the possibility of metaphysics by means of a double questioning, which bears as much on beings as on the highest being. In order to prove this, Heidegger claims that, contrary to a widespread view, the aim of critical philosophy is not to provide the laying of the ground for the modern mathematical science of nature. Mathematical physics, he argues, provides no more than an "indication" or a "direction" for the connection between ontic experience and something more fundamental—ontological knowledge.[5] Because the critique aims at providing the laying of the ground for metaphysics as a whole, not the positive sciences, it discloses the inner possibility of ontology.

This book follows and expands the negative part of Heidegger's insight concerning Kant's critique: namely, the conviction that transcendental knowledge cannot be equated with the grounding of the positive sciences. But in the process of developing this claim and doing full justice to it, we shall express reservations about the transition to ontology. Indeed, there is a key concept in the critique itself which seems to play the role that Heidegger ascribes to ontology: the concept of *totality*. Perhaps Heidegger can be accused of haste in this respect. In many ways, he has passed over Kant's concept of totality in order to articulate a connection between Kant and himself that serves his own purposes. A full clarification of Kant's concept of totality is needed before we can pronounce ourselves on whether or not ontology in Heidegger's sense is truly the ultimate horizon of Kant's *Critique*. As it turns out, the critical concept of totality is dealt with in the Transcendental Dialectic of the *Critique of Pure Reason*; the concept of world as totality is one of the three Ideas of Reason—the other two being the Idea of soul and the Idea of God. But

Heidegger, in his own book on Kant, is virtually silent on this whole section. If the Kantian expression of Heidegger's concept of Being were to be found in the Idea of totality, there would follow some dramatic consequences for our entire reevaluation of the significance of critical philosophy.

Heidegger's fundamental point of departure from Kant lies in the need to give an ontological dimension to human subjectivity. Kant is not aware of such a dimension, because his transcendental subject is a point *for which* there is being—it constitutes the phenomena of the world—but it is itself irreducible to being. Kant's transcendental subject is thus determined purely negatively, as that which is not being; the possibility of interrogating ourselves about the being of this subject is not even available. Heidegger's strategy is to bring the transcendental down to the factual. Human being must be viewed from two different standpoints: as transcendental consciousness (which is not a being, but which constitutes being), and as factual existence. Dasein is not merely a synthesis of these two structures. It is not the empirical human being, because it has the possibility of revealing the Being of all beings by means of an authentic understanding of it; and it is not consciousness in the classical sense of the term, because, as free, it breaks through being in a manner which is historical and contingent. As a whole, Dasein is historical, temporal, and contingent. Thrown into the midst of being, Dasein is not in a position to constitute being. In any constitution, being is simply assumed to escape us, and the purpose of philosophy is to make contact again with what is separated from us. Dasein is the being which has, at the same time, a comprehension of the ultimate sense of being. Heidegger's interpretation of the Kantian transcendental subject seeks to demarcate itself from the classical tradition still represented by Kant, by assuming that this transcendental subject can know only what it constitutes in accordance with its own intentions: that is, what it constructs in accordance with the categories of understanding in their empirical employment. The problem of the Being of beings would then be limited to that which is constituted. What is the reality that must be connected with the various intentions of consciousness? Kant, however, speaks of a separation between the human mind and the ultimate being of things in connection with the problem of reason in the largest possible sense. For Kant, we have a natural desire for metaphysics, because reason pushes knowledge to a point beyond any possible experience (B21); and reason will never be satisfied by simple expedients that would bring ultimate being within the purview of possible knowledge. Can this highest intention be amalgamated with all its lower expressions in connection with the knowledge of actually experienceable objects? Heidegger seems to take this unity as a matter of course, inasmuch as he does not directly address the critical concept of totality for its own sake.

There is a deep motive underlying Heidegger's stopping at the threshold of the dialectic. In *Being and Time*, Heidegger develops a fundamental ontology which is supposed to be prior to any possible metaphysics. In this fundamental ontology, the analytic of a contingent, finite, historical, and temporal place is developed—the place at which Dasein asks the question concerning the meaning of being. At the end of this analytic, we are left to wonder about this ultimate issue: how can we be sure that Dasein's interpretation of the meaning of being is the true comprehension of this meaning, not just one vision of the world among many others? All we have found in the analytic of Dasein is proof that any comprehension of the meaning of being actually starts from Dasein. But it could still be that this comprehension is a projection that fails to make contact with the sense of being. Could fundamental ontology be the preparatory step in the direction of ontology? In *Kant and the Problem of Metaphysics*, Heidegger undertakes to complete the task. He wants to show that the manner in which Dasein interprets being is also the very deployment of the meaning of being. Dasein's pre-understanding of being is not the ultimate foundation, because it is itself grounded in the inner deployment of the meaning of being. Now, Heidegger finds in Kant's *Critique of Pure Reason* the first traces of his own project of fundamental ontology. Kant's "Copernican Revolution" is based on the insight that Dasein is not just one among many beings. The metaphysical question of the Being of beings must be clarified by means of a prior examination of the actual place in which the question is raised. Heidegger disagrees only with the explicit answer provided by Kant in order to describe this place and bring it to intelligibility.

The temporality (*Zeitlichkeit*) of Dasein makes it historical. What Heidegger needs to find in Kant is a concept of transcendental subjectivity that is itself time-like, by contrast with Kant's own claim that the transcendental subject constitutes time, but is not itself temporal. Heidegger argues that transcendental imagination is, in fact, the deepest root of the critical theory of knowledge. Intuition and understanding are presented as two separate faculties. The former accounts for the presence of beings in space and time, for the fact that being is positing. The latter for the fact that we can understand these beings by means of certain mind-dependent structures (categories of understanding). Are presence and structure two different senses of being? Is there not a deeper level at which these two senses merge into one sense of being? The transcendental theory of the schematism, in which Kant articulates his doctrine of imagination, is interpreted by Heidegger as Kant's attempt to unite what was originally separated. Thanks to the transcendental imagination, the structure is given a spatiotemporal content. Transcendental imagination is the representation in time of that which is intellectual. Thus, there would be a more originary place of being than either the subject or the object. This place is indicated in Kant's theory of imagination as original time (*Urzeit*).

Dasein's interpretation of the meaning of being is not arbitrary. Rather, it is the manifestation of the meaning of being itself. The temporality of Dasein, which is described in *Being and Time* in terms of fundamental ontology, should thus allow us to take up ontology, that is, to think the meaning of Being as time. But in the Kant book, Heidegger comes up against an intermediary step separating fundamental ontology from ontology: the dialectic of Dasein, which corresponds to Kant's own Transcendental Dialectic. Between Being and time as they must appear in a philosophy that has overcome metaphysics, on the one hand, and Dasein's temporality as it appears in the existential analytic, we have Being and time as they are thought in metaphysics itself. Why does not Dasein's temporality (as pre-understanding of Being) deploy itself *immediately* in an authentic comprehension of the meaning of being? Kant's Dialectic is preoccupied with the following theme: inasmuch as reason is capable of truth, it also produces illusions which are not arbitrary (the illusions are not outright mistakes that can be corrected), but reflect the actual limits of any possible human relation to the whole of being. In the same way, in the years that followed the Kant book, more particularly in the essay "Vom Wesen der Wahrheit," Heidegger developed a metaphysics of Dasein, in which to the essence of truth is now coupled nontruth as *Unwesen*, nonessence. The revealing (*Entbergung*) of Being is concomitant with its concealment (*Verbergung*) or being covered-over (*Un-entborgenheit*). Given the rather special and pejorative sense that Heidegger ascribes to metaphysics, what he actually means is that the *Un/Wesen* of truth does not issue from the insistence/existence of Dasein, but that the latter responds to this *Un/Wesen*. The intense focus on Dasein in *Being and Time* has really been abandoned. Between the analytic of Dasein and the authentic apprehension of Being, we have the history of truth in which the truth of Being manifests itself by masking itself.

At the time of the Kant book, Heidegger has not yet developed his notion of essential un-truth. Does that mean that his stepping back before Kant's Dialectic can be excused and explained because the issue is taken up later? Certainly not. The inauthentic thought of Being is neither error nor illusion, as a transcendental mistake is for Kant, but errancy, responding to being's un-concealment. In terms of Kant's Dialectic, however, the point of errancy would be no more than a half-measure. The dialectical illusions indicate a division inherent in the totality of reason's system, the marks of which can be identified thanks to the precision of transcendantal logic. Of course, this is not to say that Heidegger does not interpret Kant's Dialectic because he cannot do so in terms of his own concepts. More seriously, in examining the transcendental logic of illusion as Kant thematizes it, we will find that any attempt to ascribe illusion to the work of the faculty of originary imagination is explicitly discarded by Kant. Furthermore, looking beyond the *Critique of Pure Reason*, we will find that, in another species of illusion (the dialectical appearances of teleological judgment), the need to refer to a faculty of understanding higher than our own (the divine understanding which is immediately intuitive, without separation between thinking and being) can only be carried out by giving up the power of imagination altogether. In both dialectics, Kant bases his conclusions on the transcendental principle of waiting—the postponing of reason's fulfillment in being. Neither the world, as an Idea of Reason, nor the divine understanding, as a projection of our own understanding beyond its capacities, finds its roots in imagination as unlimited power of revelation. Imagination opens up the synthetical activity of the mind, whereas reason postulates the possibility of enclosing, of completing a synthesis, even though the actual completion is never more than the horizon of knowing or acting. In Kant's dialectic, the residual opening present at the horizon is not the work of imagination. Put it differently, what bothers Heidegger is that Kant's metaphysics of man is an intermediary step that lasts forever. This worry must have something to do with the steadiness of Being itself.

What, then, is the critical concept of totality, and where does it take us? The task of understanding that Heidegger has bequeathed us is spelled out by Heidegger himself in the last two pages of the Kant book.[6] The content of the Transcendental Dialectic cannot be purely negative, as if Kant merely wanted to destroy past dogmatic systems by applying the results of his new position articulated in the Transcendental Aesthetics and the Transcendental Analytic. Moreover, if a positive problematic can be extracted from the dialectic, shall we not have to develop it in accordance with some presupposed infinitude? Heidegger's difficulties with Kant's dialectic are themselves rooted in earlier distortions of Kant's ultimate project, which were meant to reveal its true significance. Particularly important are the interpretations of Hegel and Nietzsche.

Hegel makes an astonishing comparison between Zeno and Kant, in order to characterize ancient dialectic as superior to modern dialectic. Kant distinguishes between a world in itself, of which we can know nothing, and the appearances that we actually know. We would then have two absolutes: the absolute infinity of the superhuman thing in itself (noumenon), and the finite absolute concealed within the human thinking subject (which Kant refers to as the originally synthetic unity of apperception). The former absolute knows of no mediation, whereas the latter puts us in a relation to what is outside ourselves. Hegel sees in this a strange double movement of exaltation and mortification. The world in itself must be absolutely true, it is only our own behavior (*Betragen*) that ruins it by clamping a mass of determinations upon it. This is the crucial difference between Zeno and Kant: whereas according to Kant the world is ruined by the spirit, Zeno argues that the world of appearance is untrue both in itself and for itself.[7] Between being and knowing, Kant chooses knowing: Knowledge dictates its conditions to being, which is now conceived as an appearance. With respect to being, knowing is both weak and false. It retrieves its strength if it is ready to limit truth to the world of appearances.

However, once this interpretation of Kant's project is accepted, there is hardly anything to stop the movement of contamination of truth by falsity, of alleged strength by real weakness. Against Kant, Nietzsche goes one step further in the reduction of being, from ontology to value, when he says that the essence of a thing is in fact no more than my *opinion* concerning the thing. Being is essentially what it is worth, that is, the authentic "es *ist*" is in fact "es *gilt*."[8] Consequently, instead of being logically opposed to one another, error becomes the basis of errancy, the presupposition of any thinking. Before there is thought, there must be poetic fiction. ("Bevor 'gedacht' wird, muss schon 'gedichtet' worden sein."[9]) This composition of the world, which takes precedence over any thinking and determines its direction from within essential errancy, has the double character of the poetic and the practical will. For life and its praxis impose their ends on both knowing and being.

Will the poetic and practical ground of life resolve the duality of absolutes pointed out by Hegel? The point is not so much to answer this question as to see what we lose when Kant's project is interpreted in terms of a tension between two absolutes. Modern neo-Kantianism has not really changed anything in our perception of Kant's achievements in terms of a duality. But the duality is not so absolute because we have to live with it. In his objections to Heidegger's interpretation of Kant, Cassirer remarks that the duality between the appearance and the thing in itself cannot be overcome by using the single framework of temporal existence. Man is not just finitude but also freedom, which is atemporal because the moral law produces its object directly, without the mediation of the temporal schemata of imagination. For Cassirer, the true opposition that strikes at the heart of critical philosophy is finite (temporal) knowledge versus freedom—what is versus what ought to be. As he puts it, when we are conscious of our freedom, "we remain entirely in ourselves and, at the same time, are in principle lifted beyond ourselves."[10] Heidegger would reply that the simultaneity involved here cannot be consistent with the temporality of Dasein's inner temporality, which, in turn, compels us to cast doubt upon the transcendental validity of a life according to atemporal principles. Are we bound to accept an impossible tension, or shall we do away completely with the duality? Kant would protest and say: Reason aims at the unconditioned, but deploying in time the movement toward the unconditioned amounts to transforming this unconditioned into an Idea of Reason. Heidegger's Being is an Idea of Reason. But of the traditional *metaphysica specialis* Kant retained three Ideas only—the soul, the world, and God. Shall we learn anything philosophically by enlarging Kant's dialectic and including Being among the Ideas? Quite apart from Heidegger's own project, apart from its more or less clear attachment to the Hegelian/Nietzschean tradition, we must begin by asking: what is Kant's justification for distinguishing, as he does, between two senses of philosophy?

What is it that remains unacceptable in the scientific construction of a worldview? For Kant, the greatest example of a philosopher who struggled to promote a similar distinction is Plato, the very founder of the tradition which sees philosophy as science. It was Plato the *academic* who, Kant writes,[11] took upon himself a task as crucial as that of the critical enterprise, namely, to elucidate the possibility of a synthetic cognition a priori. The perversion of modern times lies in the fact that what was originally an academic (scholastic) task had to define itself otherwise (e.g., philosophy as a cosmic concept), because some modern philosophers appropriated a mystical-Platonic language, borrowed from the freer style of Plato as a letter-writer. Kant's target here is Jacobi and his disciples. And his complaint is that those who speak such a language do so hermetically, because they confuse a mere clarification of the human possibilities for knowledge with an expansion of knowledge that comes close to divine understanding. They adopt Plato's comparison of the illumination of the soul viewing the sun as a mere foil for their claims that the most basic teachings of sensible experience are to be distrusted. A claim of this sort, Kant tells us, is found for example in a work by J. G. Schlosser of 1795, who used a German translation of Plato's letters in order to justify his own claims in favor of the mystical destination of philosophy. From this book, Kant cites the following statement: "All human philosophy can only depict the dawn; of the sun we can only have a presentiment." Against this quick leap into feelings that repudiate everyday life, Kant advocates a slowing down. The actual experience of the sun is both a bridle and a condition of possibility:

> But really, no one can have a presentiment of a sun if he has not already seen one; for it could very well be that on our globe day regularly followed night (as in the Mosaic story of creation) without anyone ever being able to see a sun, because of the constantly overcast sky, and all our usual business could still follow its proper course according to this alternation (of days and seasons). Nevertheless, in such circumstances a true philosopher would indeed not a have presentiment of, not *surmise,* a sun (for that is not his thing), but perhaps he could still *deliberate* about whether this phenomenon might not be explained by assuming an hypothesis of such an astronomical body, and he might thus by good luck hit on the right answer. To gaze into the sun (the suprasensible) without becoming blind may not be possible, but to see it adequately in reflection (in the reason that illuminates the soul morally) and even in a practical respect, as the older Plato did, is quite feasible.

The sun: however remote it may be for the ancients, or however close it may be for the moderns (in the wake of the Copernican revolution), remains the symbolic touchstone of the truth of human existence. Kant's designation of the seriousness of Plato's philosophy as academic, and his concomitant claim that his own critical philosophy departs from it to become cosmic, hinges on the fundamental experience of seeing a sun. The ordinary course of human affairs could certainly proceed without a vision of the sun. The true philosopher, however, soars above this ordinary course by advocating conjectures about such basic facts of experience as the days and the seasons. In principle, these conjectures could lead to admitting the existence of a sun, even if no one sees it.

The experience of the sun, though undeniably a factual experience, is thus, in the historical conditions of human life, central to human experience, in that it mandates the reversal of what originally counted as academic, and what as cosmic. Before giving himself over to moral reason and to the practical standpoint, the true philosopher must devote full attention, and therefore take the risk of becoming blind, to that particular question of nature about the sun. In doing so, he even takes the risk of blinding all of humanity; for, in contrast to the mystagogue of modern times, he takes no interest in cultivating the select adepts and initiates of his sect. The moral law already carries me higher than the sun. As Kant puts it at the end of the *Critique of Practical Reason*, there is a deep connection between two extreme visions open to us: "The starry heavens above me and the moral law within me."[12] The first vision is annihilating and makes me fall back upon the earth, which is but a spot of the universe where I have arrived without knowing how. The second vision forces me to search for my own identity in the universal and the necessary, which raises me above myself once again. The stars that shine over our everyday horizon remind us of a common enigma which traverses each of our lives. The starry heavens lead to respect for the moral law, but *between* the darkness of the night sky and the darkness of the innermost depths of my own self, there is the sun. In connecting my free reflection with that of all other people, the sun also reminds me of the authenticity of my terrestrial condition.

Kant once addressed the question of what it means to orient ourselves in thought, and his answer was preceded by an analysis of orientation in space.[13] His aim was to reach a more lucid understanding of the nature of subjectivity involved in the principle of orientation. He begins by asking: how can I orient myself in a dark room that is familiar to me? No purely conceptual relation will be of any help to me; in order to find my way, I can only rely on a lived experience of some kind. The left/right distinction will help, inasmuch as it is rooted in a subjective feeling that owes nothing to the logic of a concept; a judgment on the correct position of objects in the room will be possible, even if we imagine that everything has been rearranged in it during my absence, so that what was at the left is now at the right and vice versa. However, in order for the feeling to be awakened in the first place, Kant adds that all I need is to be able to seize on a single object whose position is present to my memory. Against Kant, Heidegger argues that in this context memory only serves the purpose of reminding us of our being-in-the-world, which is more primordially constitutive for the possibility of orientation than the feeling for right and left. [14] Has Kant been really oblivious of the world viewed as such a fundamental premise? He states quite explicitly that the dark room is known to me, just like the familiar streets of a town at night in which I walk and make the proper turns even though I am not able to see distinctly one single house. In a dark room, is not the strangeness of the situation due to the fact that I am *simultaneously* in and out of my own world? Have I not suddenly *lost* my world even though I am still physically in it? If only for that reason, the object in the dark certainly loses its property of being objective, so that my memory will be of no help. My memory has now the perplexing effect of propelling me into another world, an other order of being, at least temporarily—this is the world of impenetrable darkness, that can be compared with the immeasurable supersensible space in which the Ideas of Reason reside.[15] But Kant's example shows that this other order of being is not automatically or immediately a projection into the world of practical law.

Before taking an interest in freedom, the true philosopher must then blunder with a philosophy of nature which touches upon the cosmic meaning implied by our contact with the sun. In that sense alone can we say that scientific philosophy (the clarification of the sense of natural cognition) is also worldview philosophy (the clarification of the inescapably familiar experience of the world). For mystical feeling and its excessive haste, the true philosopher substitutes deliberation, which is to say, reflection, founded upon a *faculty* of feeling. Now, as soon as we take into account the need for a phenomenal experience, a vision of the sun, in the sense in which Kant understands it (a "not-at-home" which reminds us of our abode), we can express reservations concerning Heidegger's fundamental reversal of familiarity and homelessness: "That kind of Being-in-the-world which is tranquillized and familiar is a mode of Dasein's uncanniness, not the reverse. From an existential-ontological point of view, the 'not-at-home' must be conceived as the more primordial phenomenon."[16] Just as it invites us not to rush at the practical concerns of life, Kant's appeal to the experience each of us can have of the sun can lead us to postpone the Heideggerian reversal leading to the foundation of my being as solitary and distressed. If our immediate environment were different from what it is (if, for instance, the earth were continually shrouded by clouds), the true philosopher would have no need to reflect on himself, since the natural world would already offer him the means for his reflection: he would adopt the astronomer's patience. But everybody knows what the natural world is. The true philosopher and the astronomer would then never have anything to say to each other—they would both be at home in their own reflections—if at least each recognizes in his own terms the genuineness of the experience of the sun. Heidegger's existential-ontological point of view would be the only possible alternative, if and only if the scientific experience had itself exhausted all of its possibilities.

Quite characteristically, Heidegger interprets the fundamental project of modern mathematical physics (the so-called Galilean project) *as if it had actually succeeded*. According to this project, communication between observers and nature takes place in accordance with experimentation, which is essentially a projection of possible knowledge upon the possibilities of nature. The human project extends nature's own possibilities. The process has the peculiarity, that since the properties of nature are extended by our own mathematical reflection, they are appropriated in such a way that they stand in exact correspondence to what the knowing subject actually already had in advance, though in an indefinite way. Experimentation is the coming-to-presence of this correspondence. In *What is a Thing?*, Heidegger explains that the project thus gives credit to the mathematical as that evident aspect of things within which we are always moving and according to which we experience them as things at all, and as determinate things. Between us and the things, the dialogue does not know of any limitation as long as we retain of nature only what is calculable, that is, as long as the process is both learning (taking up) and teaching (offering). By implication, such coming-to-presence consigns the non-mathematizable features of the natural world to artifacts of the human senses. The Galilean project protects itself in advance from any refutation from the world of unmediated sensible experience.

Heidegger returns to Kant's distinction between the academic and the cosmic senses of philosophy in "Vom Wesen des Grundes." Here he distinguishes explicitly between the cosmological significance of Kant's concept and its "existential" counterpart. Existentially, Kant limits his investigations to anthropology, that is, an understanding of existence that does not go beyond a certain experience of life. Kant does not address what Heidegger takes to be the more primordial character of the phenomenon of world: the *Umwelt*, which takes the measure of human existence according to the affective tonality of its being given over to a world into which it has been cast. This tonality comes to expression in the one genuinely encompassing concept that must be admitted at the basis of all experience, namely, that of care (*Sorge*). Heidegger says explicitly that nature (in the sense of the sciences of nature, but also in a quite primitive sense) is not included in this investigation of Dasein's existence, because care is merely the basis for the problem of nature. Once again, the question arises whether such an investigation could ever refashion the contact with nature that is given in mediate (scientific) or immediate (primitive) experience. For Dasein, indeed, the totality is truly understood prior to any explicit grasp of the whole of beings in its specific interconnections, its regions and stratifications. Of this totality there remains, therefore, only a horizon that has lost the phenomenal diversity of the ontic experience of the world. To return to Kant, it is obviously not to be taken for granted that philosophy, in its cosmic expression, should make contact with the full diversity of the whole of nature, given that the limiting experience of the sun draws us into a dangerous game of oscillating between the beyond and the below of any possible experience. But in any event, the horizon of freedom does not bring about a shattering of the suspect attachment of transcendental philosophy to the model provided by nature. Before passing to the practical mode (or the existential mode, in the sense of Heidegger), a cosmic philosophy must involve itself to the farthest possible extent in the ontic meaning of natural objects. This ontic meaning is, indeed, the place of an authentic opening unto the world. Yet once the leap into freedom (or existence) is effected, this sense is perhaps effaced prior to having been really understood.

As against the separation of transcendental philosophy from the scientific philosophy of nature, Schelling is the philosopher who has perhaps most lucidly grasped the need for taking up the Kantian project once again. In his *System of Transcendental Idealism*, he explains that—for transcendental philosophy no less than for any science of nature—the unconditioned (*das Unbedingte*) can never be identified with any thing (*Ding*).[17] Both disciplines are worried about the same origin. For matter is nothing originary, but is appearance (*Schein*) for natural science no less than for transcendental philosophy. However, Schelling goes on to argue that transcendental philosophy and philosophy (science) of nature relate to the *same* totality of knowledge, which they approach in opposite directions. The movement from the object (nature) to the subject (self-consciousness) develops, in its own manner, the very same determinations that transcendental analysis unfolds from a fundamental act of unconditional subjectivity which opposes itself to the world. Schelling would like us to believe that the unconditioned of transcendental philosophy and the unconditioned of the philosophy of nature cannot be distinguished in their being. In the final analysis they must be absolutely identical, because they are both equally unknowable. In this way, Kant's actual project is covered up again at the very moment of being reenacted. For why should the unknowable be a point of convergence? Kant's sense of the unknowable does not bring together; it only indicates possible paths in the direction of bringing together. Just as Heidegger's reversal cannot really begin unless we could be sure that natural philosophy has given all that it could, so the absolute aimed at in Schelling's philosophy of identity is perhaps *not yet* a place of reconciliation of opposites.

In his own commentary on Schelling, Heidegger has pointed out the fundamental precondition for philosophy today: "To note today that Kant did not yet know anything about possible historical transformations of the use of reason is not difficult after a century and a half of historical, anthropological, and psychological research; it is simply tiresome and unfruitful. However, it is difficult to raise our task and work up to the level of Kant's thinking again."[18] Strikingly enough, Heidegger does not include the most obvious of these transformations that make us falsely believe that Kant's philosophy has been overcome: the developments of natural science. Heidegger's omission, whether deliberate or accidental, is fraught with remarkable consequences. That Kant's transcendental investigation of natural science is superseded by post-Newtonian science is obvious. However, post-Newtonian science is perhaps completely alien to a purported transcendental foundation. Is this a sign that science has come to an end, that the scientific worldview has exhausted all its possibilities? Perhaps it is the other way around. In disentangling itself deliberately from the yoke of transcendental foundation, science has turned in the direction of an entirely new kind of foundation, very little of which is understood at present. If this latter possibility cannot be simply ruled out, then however historically determined it may be, and precisely because it is the last attempt in modern history to understand science absolutely, Kant's investigation remains today the ultimate touchstone for the clarification of *any* possible natural knowledge, and thus also for the knowledge of the whole in the context of modern cosmic philosophy.

Nature and Freedom

In Kant's critical philosophy, the distancing from nature, achieved by the true philosopher on the way to freedom, takes place in the third cosmological antinomy of pure reason. The process reflects the painful extraction accomplished by reflection, as a result of which, in the antinomy, transcendental freedom is assigned to the idea of world. Yet, at the same time, as a foundation for practical philosophy, it liberates itself from the world. Freedom emerges from determinate being, instead of being simply suspended in it. If it were true that freedom is a property of the faculty of desire, the will of a reasonable being, then it should have belonged to the psychological idea which deals with the "I" of "I think." Since this is not the case (A448/B476), freedom in the Kantian sense cannot be said to belong to the essence of man. The ability of human beings to choose between right and wrong, good and evil, does not proceed from their freedom. Rather, freedom is the beginning of a series which insinuates itself into the causal series of the world of experience. It is to be met with only in acting, and it works as the principle of action. Kant expresses this fact by distinguishing two "characters" in the series of the sensible world, that is to say, two modes of causality. One is empirical, according to which any cause is the effect of an antecedent cause; the other is intelligible, as a power absolutely to initiate a series of causes. (A539–41/B567–69). These two characters do not define two worlds, because the intelligible character does not at all belong to a higher understanding. That is why transcendental freedom is still part of the cosmic concept of philosophy. There is only one world, namely, the sensible world, in which these two orders manifest themselves in accordance with their own principles, like two voices of the same melodic line. As Heidegger puts it, "freedom is nothing else than natural causality thought absolutely."[19] But when it is thus thought through, nature contradicts itself (its two modes of causality degenerate into a conflict between thesis and antithesis), precisely because we can never think absolutely. A bold stroke is needed in order to make it thinkable. Kant therefore invites us to think about the unity of irreducible terms, a unity such that the two terms *remain* opposed to one another (even though the thesis and the antithesis are both true)—the two voices are always discordant, there is no tonality of being that could ever fix the melody of the world in anybody's ear. As it turns out, thinking through such a unity also requires that we move away from the cosmic concept, or that we swing over to another horizon for the interpretation of human freedom as the *absolute* beginning of a series.

Kant distinguishes between two senses of the unconditioned: the mathematical and the dynamical. To the former belongs the cosmic concept "in the narrower sense" (A419–420/B447–48) which deals with "the world of the great and the small." This employment of the cosmic concept covers the idea of world, which signifies "the mathematical sum-total of all appearances" (the first two cosmological antinomies). But when the world is viewed as a dynamical whole, it must be called nature, inasmuch as it concerns the unity in the existence of the appearances, not merely their aggregation in space and time. In contrast to the cosmic concept, Kant calls the idea of such a dynamical whole the *transcendent concept of nature*. Freedom, as unconditioned causality in the field of appearance, belongs to the latter class of concepts. But whereas world and nature are not opposed as irreducible elements (they indicate the same reality), freedom and nature *are* opposed to one another in respect of the two above-mentioned orders, namely, the empirical and the intelligible. Nature, as a dynamical whole, produces its own series of events within those of the mathematically homogeneous series. Freedom organizes its series by using the same substrate as does dynamical nature, but it is another dynamics altogether, in which "change" requires no "dynamical determination in time, and therefore no causal dependence upon appearances" (A541/B569). What is an event that appears in the sensible world, and is yet wrenched away from any temporal relation to what precedes it? Human being must first have, and show, the *strength* that such a wrenching presupposes. That is, each of its resolutions, which testifies to its potentiality for free action must be accompanied by a resolution of a different order: the power to manifest that he is *alien* to the world in which s/he nonetheless continues to live—the mechanical world of sensible events, which s/he has succeeded in making intelligible by means of categories and principles of understanding. The third antinomy gives him or her the tools for this wrenching.

Heidegger realizes that the exposition of the conflict between thesis and antithesis in the third antinomy results from the fact that the problem of freedom belongs to the problem of the world.[20] What motivates the search for a resolution of the conflict between causality through freedom, and causality according to nature, which is to say the possibility of union between partners in conflict? Heidegger does not see here any forceful innovation. According to his reading of Kant, this possibility is simply the result of borrowing from the reflections of ordinary reason. For natural experience already presents to us a unity of laws. We never leave the world, because man, the plain "material evidence"[21] of the world, is the connecting link between the two causalities. To think the possibility of a union between nature and freedom amounts to thinking "the possibility of man as being-in-the-world."[22] Insofar as he integrates the problem of man into the cosmological problem, Kant also would have reduced nature to something simply "selbstverständlich." But to think through natural causality, to attain its ultimate principle, will this amount, in the end, to imitating its surface, as it is immediately given to ordinary reflection? What is the world that grounds the Kantian Being-in-the-world, since at the very moment when the powers of natural causality give out, the cosmic concept has already burst forth in the transcendent concept of nature? What is presented in the third cosmological antinomy must be already the outcome of a thought which clarifies a motivation, deeper than common reflection, for engaging in dialogue with nature. This is a dialogue in which the cosmic sense of philosophy is critical.

Being and Knowing

Being and knowing: as soon as I think the relation of one term to the other, I am deprived of my self-possession. I cannot, indeed, hope to know anything whatever about being unless I am prepared to lose myself in it, to abandon the prerogatives of my own existence. Yet I who think am always still here. Unless I renounce knowledge, thus respecting being without marking it, in whatever way, by a relation of knowledge, I can only try to regain possession of myself. To recapture what I always already had, not to let myself be intimidated by the suctioning force of being: such is the first manifestation of that resistance which constitutes the freedom of a being that knows. I lay claim then to what I owe to existence. Yet my request encounters no echo whatever in being. Since it encounters nothing other than its own prerogatives, it encounters nothing that could delimit it. Knowledge is the development of that freedom which owes its force only to a lack of resistance on the part of being, to the lack of interest, on the part of the world of things, in my miserable interrogation. Sometimes this force fancies falsely that it encounters obstacles on its path. It quite deliberately invents obstacles—the concept, the sensation, being—so as to test and gain the assurance of its power. These obstacles it surmounts more easily the more they pose themselves *subsequent* to its confident outset: the free force finds them only in looking back behind itself. Kant expresses this movement with remarkable lucidity: "it is, indeed, the common fate of human reason to complete its speculative structures as speedily as may be, and only afterwards to enquire whether the foundations are reliable" (A5/B9). Consequently, the frail force of a light dove sufficed to allow freedom its flight.

If, however, it were to be confirmed that the invented obstacles are not, in the end, all that different from the density of the world in its own being, the test whereby knowledge secures its acquisitions would become a source of irremissible disquiet. This is exactly what happens once philosophy discovers its critical project. Critical vigilance discovers the dogmatism, and the naïve arbitrariness, of the spontaneity of freedom that is taken for granted, the spontaneity which pushes us forward, which enables us to face and surmount obstacles. It tries to exercize its freedom in the direction which dogmatism tries hard to flee by means of its customary ruse: regression. It endeavors, at every turn, to return to the origin of the arbitrary dogmatism, to retrace the steps of dogmatic confidence and delay its claims, because dogmatism is answerable also for being, and not merely for its own ruse with false obstacles. This would lead to an infinite regress if, for the last time, the regress itself had to play games with being, and flee it, so as to come back to the zero point where knowing faces up to being. Is the regress, then, still an exercise of freedom? Such is the question Kant poses in the Transcendental Dialectic of the *Critique of Pure Reason*, where he describes the retracing of dogmatic steps as the infinite regression brought about by human reason, which passes from the conditions of what exists to the latter's ultimate condition that is itself unconditioned. Human reason, of its own nature, becomes entangled in a conflict with itself because, trusting its own naïve spontaneity, it fails to notice that, at one moment at least, it comes up against being. It goes on to pretend—by a process that Kant calls "transcendental subreption"—that it does not recognize being because, however much being may come toward us without our having incited it to do so, it will, in any case, be unrecognizable. The shock of the encounter, nevertheless, leaves its trace.

For, in order to prepare itself for the shock and to protect itself against the consequences thereof, reason fashions its own weapons, so as to fight on undisputedly equal terms with being. These weapons are the interminable polemics (called "antinomies') which turn being away from the field of reason, leaving reason to imagine that it contemplates only itself. In the antinomies themselves, the delaying tactic of critical reason, against the rush forward demanded by dogmatic reason, has become far too intrusive. A critical solution to the conflict is needed, which will enable us to advance without the undesirable haste of dogmatism. But what is it that such a solution will actually solve, since no magic will ever remove the primordial disquiet from which the critique nurtures itself?

There is still another manner of distorting the primordial disquiet of knowledge as it grapples with being. It is a matter, here, of deferring the disquiet by passing from knowledge to ethics. Ethics installs itself, from the outset, in the unconditioned that theoretical reason so desperately aims at. It can do so either with perfect calmness, as if none of the effort of reason were to show through, or else at the cost of a new ruse, masquerading as the authentic accomplishment of that effort. However, whether it is immediately within truth, or whether it dupes us so as to achieve its ends, ethics, rather than continuing to place full confidence in the exercise of freedom, calls this very exercise into question. It questions itself concerning the value and merit of the initial questioning of being on the part of knowing. The contact with the world is no longer what unsettles me, but rather my own responsibility in the face of the implications, for life, of posing the question.

The so-called Copernican revolution that Kant brings about, so as to carry out the critical project, has the peculiar character of being an exercise of freedom of which the meaning, from the outset, is to call freedom into question. Kant achieves this thanks to his distinction between *two kinds* of being, the appearance and the thing in itself. In Kant's own terms, it is a matter of experimenting with the fundamental concepts of all thought. The experimentation calls for a staging and for a proof. To the two kinds of being there correspond two proofs supporting the validity of the hypothesis. To begin with, there is a direct proof that faces up to being without guile. For being is, from the outset, under its control—it is appearance, the own offspring of the proof. But this land of truth, as Kant calls it, is "surrounded by a wide and stormy ocean, the native home of illusion, where many a fog bank and many a swiftly melting iceberg give the deceptive appearance of farther shores" (A235/B294–95). I must admit that the appearances may have been engendered by being, rather than by any interaction with myself. I cannot, consequently, avoid seeking an indirect (dialectical) proof, where reason loses control of appearances, its own offspring, and calls being to account. Not that it is a matter of seeking reassurance. It is a matter, rather, of exacerbating once and for all the primordial agony of my position within being, of engendering illusions as one engenders truth. There are no antinomies of dogmatic reason, which would be superseded by the antinomies of critical reason, but rather a single dialectic that allows itself to be manipulated in keeping with the interests of reason. Kant therefore thinks that the indirect proof reaches its term if one succeeds in thinking the thing in itself as *the other* of appearance, as if it were the latter's back side seen in a mirror. In this way, since I have compelled illusion to speak my language, that which escapes me—the offspring of being—is after all still in my power. In thinking freedom by the same movement of thought which puts it into question, *the critical project reaches its highest point just before consciousness founders in ethics properly speaking*, which is to say in the reflection on my possible inability to master what escapes me.

In this way, if this highest point could be articulated for its own sake, the rush for new demands that cannot be fulfilled by the critique is postponed. The critical project should not—in the sense of not yet—be overcome by abrogating the original duality of its absolutes. Something must be done to change Hegel's verdict on Kant, as interpreted by Heidegger, according to which critical philosophy wants simultaneously to reach the absolute and to get by without *the* Absolute. For the desire of the absolute is defined by Kant as *natural*, it falls short of the properly *critical* desire yet to be characterized. Following Hegel, Heidegger writes: "The seemingly critical fear of rash error is really the uncritical evasion of the truth which is already gathered there."[23]. The critique *is* an evasion, which is all the more enormous because it can accept no truth (in particular, that of Being) which is always already deployed in some way before us. Nor should we be asked to live in the impossible tension between what is and what ought to be. The examination of the highest point requires that we refrain from any prejudice concerning the transcendental or metaphysical nature of either term. This kind of examination puts us on the track of an authentically phenomenological reading of Kant's dialectic. Indeed, a prejudice shared by commentators of the dialectic is well reflected in Cassirer's interpretation. The transition from the concepts of understanding to the concepts of reason, from the Transcendental Analytic to the Transcendental Dialectic, is such that the latter concepts, as he puts it, "never refer immediately to intuition but rather pertain to the use of the understanding itself, to which they intend to give the greatest systematic unity."[24] If the sense of the unconditioned aimed at by reason were already fixed by the limits of objectivity constitutive of the understanding in its empirical employment, then reason would have nothing to learn from its own indirect relation to intuition. But if this indirect relation resulted from reason's ability to neutralize the action of the understanding in certain well-definable circumstances, and put it out of play, then the non-immediacy of its relation to intuition would be the vision of the otherwise concealed essence of the appearance. When the understanding understands something which is given immediately in intuition, Kant tells us in the Analytic, it recognizes, that is, it sees itself at work in the ordering of the immediately given manifold. When I know something, I also know myself in this something inasmuch as a trace of my own intellectual activity is visible in it. Thus, the essence of the appearance, or the appearance *as appearance* and no more than appearance, is concealed at all times, precisely because it can only be recognized. Our mental appropriation of the appearance has the effect of splitting the cognitive powers into pure sensibility and pure understanding, whereas we still would like the appearance to be redoubled (i.e., to appear as appearance) so that the world itself could teach us how to make sure that it is what it is. As we move to the Dialectic, we get into the field of illusion, reason begins a discourse about things without even noticing that it speaks about unknowable things in themselves. But how can we pretend to see what cannot be seen, let alone speak about it? We are led to suspect that the non-immediate relation to intuition is another kind of concealment, a postponement of immediacy which enables us to see the appearance as if we were in it, without any need for self-recognition. There is a *phenomenality of the world*, namely, the spatiotemporal level of immediately given experience which gives objects their prior phenomenal character; but any attempt to capture the appearance as no more than appearing only leads to a new phenomenality, just as irreducible as the first. This is the transcendantal appearance, understood as spontaneous production of thought. On balance, a meaningful dialogue between thought and the world is possible only through the intermediary of a play of illusions proper to the *phenomenality of thought* itself, since Kant names "principles of *intuition*" those regulative (dynamical) principles of reason which (unlike the mathematical principles) fail to be constitutive (A664/B692).

Consequently, by privileging the *faculty* of imagination at the expense of intuition and concept as two irreducible poles, Heidegger's interpretation leads to the unacceptable view according to which the "phenomenon," whether an *Erscheinung* or *Schein*, finally appears in the same way. Instead of interpreting any regressive movement toward ultimate conditions of possibility as an opening toward a more originary domain, we will have to interrogate the regression as if it were an originary domain to itself. But before we can engage ourselves in this phenomenological exercise, we must ask whether the broader framework of questioning within critical limits is still legible once the Transcendental Dialectic is read as a phenomenological moment. What is it a moment of?

In his lectures on logic, Kant says that, when taken in its cosmopolitical sense, philosophy faces the following four questions: What can I know? What should I do? What may I hope? What is the human being? The connection between these four questions reflects the innermost interest of human reason. Does not our interpretation demand that we focus exclusively on the conditions allowing us to pass from the first to the second question? But if that is so, do we not run the risk of depriving ourselves of any means to approach the one ultimate question—the question of hope, which opens up the realm of what is authentically human in the human being? Commenting on the questions of ability, duty, and hope, Heidegger, in the absence of a positive interpretation of Kant's Dialectic, believes that any possible answer to them can only ratify the essential finitude and negativity of the Kantian ground-laying. Where an ability is to be delimited, it already places itself within a dis-ability; where a duty is questionable for a creature, this creature worries about what it should not do; and finally, "where an allowing to hope becomes questionable, . . . what is asked about is what can be placed in the expectation and what cannot."[25] Clearly, if a positive account of the Dialectic is thinkable in Kant's own terms, we must prepare ourselves to find space within the critical bounds for an infinite expectation—an infinity which can sustain the paradox of being simultaneously fulfilled and unfulfilled. Again, this space is found outside the realm of ethics properly speaking.

Indeed, the moment just before ethics takes over is the point at which the critical concept that is itself the most exalted, that of totality, redoubles itself. On the one hand, we have a totality of pure nature, which incorporates the entire field of experience, however far it may extend. This is the infinite climb, where the cause through which all things happen is itself caused. But on the other hand, this totality fails to encompass the "absolute totality of conditions" (A533/B561), which contains the reason for the causal relation—the hypothetical point at which the infinite climb stops. The idea of a spontaneity which can begin to act on its own makes up for this failure, and even though the empty space between the two totalities indicates my finitude, the free spontaneity fills it. The totality of pure nature, which lacked some of its conditions, is filled up with this spontaneity and becomes the absolute totality of conditions. But the free action which supplies the missing conditions does not belong to either totality. This freedom is transcendental. Its action operates without incurring any debt with respect to one of the two totalities. Not belonging to any totality, it is pure indeterminacy. It furnishes reason the means for this forward flight before the implications of an encounter with being. And it justifies appearance.

But just as one can speak of two totalities, freedom also redoubles itself. For there is a practical freedom, the domain of ethics properly so-called. Being incapable of rendering transcendental freedom determinate, practical freedom renders it at least sensible in its own being as indeterminate. In its encounter with unknowable being, it leaves a mark of absence, namely, that if something did not come to be, that thing ought to have or could have come about. Any moral law is based on such a retrieval of absence; the moral intention cannot be caused by a material content. How, then, can it conform itself to a content without being caused by it? The work of practical freedom would not be possible if freedom itself were not double: an external versus an internal freedom. An act is moral when it is done out of respect for the moral law. The same act is simply legal when, viewed from the outside, it conforms itself to morality. It can then be done by social coercion, or by fear of punishment. The function of the state in civil society will then be conceived by Kant in such a way that, through natural causality, the content of acts is legal. For the moral intention finds in civil society an adequate content which, however, is not caused by it. Contrary to a common interpretation, Kant's doctrine of right does not derive from morality. Together with the trancendental illusion in cognition, the infinite expectation that we await in and from society is thus the other highest point of the critical enterprise, just before ethics takes over.

To the two forces of freedom there correspond two forces of knowledge, understanding and reason, which are no less opposed to one another. So as to seize upon the key moment of the critique, it is important to articulate what produces their opposition. Not that the critique could, on this point, be converted into a simple exercise in the foundations of science. The first critical experimentation with the concepts, which posits the division of the totality into appearances and things in themselves, already does not simply translate the revolution of thought brought about in science by Copernicus, Galileo, or Newton into philosophical terminology. Kant, in fact, gives us to understand that the transcendental philosopher, insofar as s/he separates the appearances from the things in themselves, is comparable to a chemist; and elsewhere he complains that chemistry is not truly a science since it lacks the mathematical foundation proper to every science worthy of its name. Despite his interest in chemistry in the *Opus Postumum*, certainly he does not expect that imminent developments in chemistry will allow the latter to sanction the scientific character of the critical enterprise. If it is true—as Heidegger holds—that, for Kant, the decisive orientation for resolving the antinomies is provided by nature, one must take into account that the relationship of the critique to nature is not simply an analogical reflection of the labor of science. On the contrary, the distance between the critique and the sciences allows us to put the Galilean project out of play. The vindication of scientific philosophy in its critical sense does not require one to regard this project as having always already succeeded.

The antinomy, moreover, is the result of a particular type of analogy (causality) that is pushed too far, that is, the analogy of experience which becomes trans-empirical. In relying upon the empirical synthesis as a model, reason seeks the unity of a synthesis which surpasses the field of experience. It ends up by articulating a redoubling of finitude. Whereas every limit

(*Schranke*) presupposes something real that encloses or circumscribes, the boundary (*Grenze*), by contrast, is a simple negation without relation to anything else. Whereas the boundary is reassuring, because it makes us believe in the existence of a definitive wall, it is the very nature of the limit ceaselessly to renew the disquiet at the very heart of the being-ready-to-hand of our factical existence. The chasm between the two senses of finitude is precisely what swallows up the Kantian antinomies.

CHAPTER 2

The Mathematical Dream of Philosophy

"I See the Trace of a Man"

Heidegger argues that, in order for thinking to move away from the "strange mistake" according to which it must be subordinated to exact science, the dignity of philosophy must be retrieved, a dignity which lies entirely in a questioning (*fragend-denken*) that takes us back to a point of contact with the unique source of being.[1] Kant, on the contrary, indicates that this dignity is in fact present in the way we come to grips with our primordially unreflected passage from one order of being to another—from the moral to the intellectual, or from the sensible to the intelligible.

Indeed, it is a matter of its "peculiar dignity," according to Kant, that philosophy follows out the paradoxical consequences of its attachment to experience as an ultimate touchstone. In reference to the laws of nature, "experience supplies the rules and is the source of truth." In the case of moral laws, experience is no more than "the mother of illusion" (A318–19/B375). Two totally different spheres of philosophical reflection open up: "what is done" versus "what ought to be done." The dangerous illusion that lies in wait for one indulging in such reflections is the ever tempting inclination to derive the latter from the former. The moral law owes nothing to the sensible, yet it is to the sensible that it refers and in the sensible that it finds the conditions of its reality. What Kant finds most disturbing is that whether one confines oneself within the moral law or attempts to derive obligation from the data of experience, one is subject to various types of illusion that result from the overlap between the moral law and experience. Indeed, Kant views the authentic philosopher as one who cannot keep himself from climbing step by step from the physical to the moral order. The first philosopher to have undertaken this climb in a manner worthy of respect is Plato, who discovered that it is not just the efficient causality of human actions that points to the need of postulating ideas as ideal models. Plato's breakthrough was essentially this: The sheer visibility of a plant, an animal, or indeed the spectacle of the heavens is enough to reveal that they all must be possible in accordance with such ideas. Plato's attempt, Kant goes on to argue, survives the "ridicule" to which it is exposed because it is worth one's effort to imitate it. Transcendental philosophy makes an effort to place Plato's philosophy in a light true to modern times.

Because he adheres to a double illusion (of morality caught in the contingencies of ordinary life, and of philosophical reflection striving to maintain its grip on the necessity of natural laws), Plato is, for Kant, both the supreme model and an object of hatred. He was "as good a mathematician as he was a philosopher" and also "the father of mystical exaltation."[2] The relationship between Kant and Plato turns out to be similar to that between philosophy and mathematics. Mathematics, viewed from the standpoint of transcendental critique, is incapable of wrestling with humanity's most profound questions—those questions that, precisely because of their ultimate character, lie at the basis of the many ways humanity sets highest ends for itself: whether the world has a beginning, whether my soul is immortal, whether I am genuinely free, whether everything has a supreme cause. Nevertheless, mathematics enjoys an indisputable dignity compared to more transitory human achievements. Some have seen mathematics as a guide for philosophy because it rises "to a degree of insight far beyond what any philosophy based on ordinary experience would lead us to expect" (A464/B492). Why would means Kant regarded as deficient answer humanity's highest concerns in ways that pass over people's ordinary experience? To re-actualize Plato's philosophy in modern times, should one try to transcend ordinary experience, or must the high goals of mathematics be lowered?

Kant writes that Plato greeted the discovery of mathematics with such enthusiasm that it lifted him above empirical concepts to a world of Ideas (CJ 363). While doing mathematics, we seem to endow our souls with the power to draw the harmony of beings from their own supersensible principle, and yet we never lose contact with immediately sensible humanity because, Kant goes on to argue, "the properties of numbers" are precisely that "with which the mind plays in music." Plato thought that the amplification of world experience mathematics provided could be explained only as a result of an intellectual communion between ourselves and the ultimate origin of all beings. This blind faith in the power of mathematics has had remarkable consequences. In modern times, it became the basis of our great admiration of nature—Platonism has been revived in modern physical science. Kant warns that this admiration of nature results from a misunderstanding and that it has gradually increased to become an extravagance (or fanaticism: *Schwärmerei*). To be sure, the encounter of mathematics and music under the aegis of intellectual playfulness is the powerful motive of the need, obscurely but vividly felt, for reconciling intellection and emotion. But mathematics enjoys a proximity with the primary experience of music which, however, comes to be suddenly abrogated when confronted with the finished product. Indeed, mathematics, Kant acknowledges, is undoubtedly a condition of the possibility of the impressions that come with musical experience; but since music merely plays with sensations, mathematics cannot contribute "the slightest part to the charm and mental agitation that music produces" (CJ 329).

Enthusiasm turned into extravagance: this is Kant's judgment of modernity. He compares enthusiasm to madness (*Wahnsinn*), and extravagance to insanity (CJ 275). Both derange the imagination: enthusiasm leaves imagination unchained, as if it were no longer hampered by laws, while extravagance goes a step further and leaves it without laws at all. Extravagance forces us to go beyond the limits of possible experience and is thus the prototype of thinking in precritical philosophy. The whole history of Western ideas progresses from the unchained to the unruled. In modern times, our extravagance lies in confusing mathematical and philosophical knowledge, or in making a deceptive analogy between the two. In his classification of the possible cases of delirium,[3] Kant defines mania (*Wahnwitz,* which he sees as comparable to extravagance) as a case of analogical delirium: "the mind is charmed by analogies, which it confuses with the concepts of similar things."

The notion that Kant uses to separate mathematical from philosophical knowledge in his own transcendental terms is itself on the verge of succumbing to that charm. Indeed, the distance between them looks so thin: he sees the former as knowledge gained by constructing concepts, while the latter is more simply knowledge by concepts. How does philosophy mature, and how do transcendental philosophers emancipate themselves from the tempting mirages of mathematical enthusiasm by dropping the demand for construction from its premises? The concept of construction, inasmuch as it meddles so closely in the affairs of philosophical knowledge—even if it is only to withdraw from it—, should come into relief as the sign that humanity in modern times remains in partial bondage to cognitive demands or ideals that we can never adequately fulfill. Modern philosophy has reenacted the search for ultimate conditions of possibility, but by lowering its object from unimpeded constructions of the mind to the actual "unfolding of the experience wherein [concepts] are encountered" (A94/B126), it has come to face a double infirmity. (1) All of our concepts need to be "realized," but only their intuitive or sensible correlates (whether a priori or empirical in nature) have the authority to do this. (2) Our minds are not directly intuitive, that is, they are incapable of generating at will the sensible manifold in reference to which our concepts can be "realized." Kant's move embodies the tragedy of modern philosophy because, in facing such a humble task, modern philosophy began by falling prey to extravagance. Indeed, Kant believes that Locke was responsible for inconsistently opening "a wide door to enthusiasm" (B128), because he attempted to derive these concepts needed in order to ground experience from actual experience itself, as if our encounter with experience could remain uncontaminated by the extraction of concepts from it. Locke's operation, then, fails to unfold experience from below upward; instead, it forms the starting point of construction above experience. The paradigm of constructibility espoused by Locke can thus be formulated in the following way: An original experience of the world remains unspoiled even when it is crushed by the weight of reflection above it. The paradigm lies at the basis of all blind attempts to reach knowledge which transcends the limits of experience.

But concepts derived from experience must share the contingency of experience and fall short of any necessity. The only way of *deepening* the connection between intellect and experience, Kant suggests, leads to a complete *reversal* of the strategy: we should be able to derive concepts from "their *original* relation to *possible* experience" (A94/B126–17, emphasis added). The new task is extraordinarily daunting, for at first sight it leaves us in a state of strange perplexity. Does not mathematics represent the highest model of necessity ever to arise from the human mind? By contrast with the philosophical project of the unfolding of experience, mathematics as a condition of possibility, Kant told us in reference to the example of music, did not play the slightest part in the finished, sensible product. On the other hand, Kant argues explicitly that the sheer fact of pure mathematics seems to be enough to prove the existence in us of a priori knowledge, independent of any empirical derivation. But is not contingency the very touchstone which testifies to our radically human and finite condition? If that is so, the demand for necessity in the philosophical knowledge of experience looks like a new, impossible extravagance, perhaps even worse than the grafting of mathematical universality and necessity upon the contingent nexus of ordinary experience. In the possibility of experience as envisaged by Kant, what is it that might conceal the key to a necessity which remains within the limits of the human horizon? It would seem difficult, if not extravagant indeed, to avoid the view that the double infirmity diagnosed by Kant is simply a consequence of the philosopher's continuing unhealthy yearning for the constructive aptitude of the mathematician. Is not the promise held out by mathematical construction the paradigm that Kant follows in all his attempts to come to grips with the "realizability" of our concepts? The only difference with the precritical glorification of mathematics seems to lie in the fact that Kant, with respect to the manifold of experience, requests a constructibility from below upward, rather than from above downward. Without this new type of construction, the manifold, so Kant tells us, could not be really unfolded. Whatever answer may be given to this question of internal logic in Kant's critical work, the transcendental philosopher finds himself in the uncomfortable position of defending the critical standpoint while becoming aware of the risks involved in postulating a sense of (non-supersensible) humanity for which no concept can be constructed.

The risk inherent in critical philosophy is to be incurred in the following way. Kant describes his so-called Copernican revolution in metaphysics, which he believes is modeled on the past success of mathematics and mathematical physics, in the following terms: What we learn a priori from nature is only what we ourselves have already inserted into nature (Bxviii). But what is the point of learning what we have already inserted into nature, since this comes from us? Whence this extravagant claim of the moderns to become as masters and possessors of nature, since nature seems to do no more than teach us how to recognize those powers that belong to us? If we want to see mathematical learning as the paradigmatic process that sustains this notion of a priori, we will be extremely disappointed. For Kant speaks of an essential difference between philosophical and mathematical knowledge. The former, he argues, considers the particular only in the universal, while the latter considers the universal in the particular (A714/B742). For instance, this single triangle which I draw on a piece of paper is still determined by certain universal conditions of construction pertaining to reason. In other words, what distinguishes the two kinds of knowledge through reason "does not depend on the difference of their material or objects." Their essential difference lies in a difference of direction (or sense) in any intellectual journey attempting to reduce the separation of the universal from the particular. The course from the particular to the universal is what Kant calls *reflection*, whereas the inverse course from the universal to the particular is *determination* (CJ 211, first introduction). In the determinative judgment, we subsume given empirical presentations under a given concept (which is thereby determined); the concept is therefore itself the principle of such judgment. But in the reflective judgment, the universal is not given: it has to be found in order to make the concept possible. Insofar as the parallelism between reflective/determinative and particular/universal can be maintained, we are led to this first conclusion: whereas in mathematics the construction of concept is also that by which the concept is made possible, in philosophy knowledge by concepts which are already given is meant to make experience possible—that is, "unfold" it. If that is so, the charming temptation of an analogy between philosophy and mathematics is ultimately completely exorcised, in the sense that mathematics can and must be learned, while we cannot learn philosophy but we can rather only learn to philosophize (A837/B805).

And yet, we cannot help but ask the immediately following question: What, then, is the principle of reflection? In posing this question we perform an act of faith that, in order to understand mathematical construction at the level of its possibility, one must henceforward situate it in a purely philosophical context. Is there something like a "first reflection" which serves as a principle of all subsequent reflections? But how could we possibly reflect on the basis of a first reflection, since we would already need the principle of this reflection? It seems that we know that with which we begin only in the case of determinative judgment.

Consider the major conclusion of Kant's Copernican revolution, namely, that we have a priori knowledge of objects only inasmuch as they conform to the constitution of our faculty of intuition. Kant asks us to see it as resulting from an experiment with concepts. Suppose that objects can be seen from a twin perspective, first as objects of experience and then as objects of pure thought (Bxviii, note). We find that in the latter case reason is involved in an unavoidable conflict with itself, which makes the twin perspective untenable because this self-conflict (antinomy of pure reason) is precisely what impedes the mind in its attempt to extend its rational conquests beyond the territory of objects of experience. In the terms of precritical philosophy, it is just the opposite which is expected: inasmuch as it corresponds to a concept taken in isolation from anything else, the object of pure thought should be free of contradiction, irrespective of whether it has or fails to have a connection with sensible experience. Kant's experiment is meant to show that the world of pure thought and the world of experience enjoy an affinity that could not be brought out, except by constraining objects in some way. Instead of leaving open the option between two ways of relating to the objects, we have acquired the notion of "concepts of objects," that is, concepts now have a relation to objects which makes the latter "built in" the former. But the price to pay in order for thought to constrain the manifold of experience is that its own self-contradiction in the case of objects beyond experience reveals just how much it itself is not free. Kant writes: "When judgment *determines,* it has no principles of its own that form the basis for *concepts of objects*. It is not autonomous; for it only *subsumes* under laws or concepts that are given it as principles" (CJ 385). Thus, in order for the experiment with concepts to be carried out and completed, we must assume that, at the very least, something like a "first reflection" is always available in the case of determinative judgments. With concepts serving as principles, determinative judgments relinquish the freedom they acquired by constraining the objects, only to avoid the danger of being exposed to a conflict with their own principles. In the antinomy of pure reason, the self-conflict of reason only echoes a dissociation within the object; had pure reason not collapsed concepts and principles of these concepts, something much worse would result, since pure reason would then "come to be at variance with itself."

Kant speaks of the determinative employment of the faculty of judgment in its own reflection (CJ 212, first introduction): this is nothing other than transcendental schematism. This method, as Kant calls it, exempts us from seeking the rule of judgment outside our capacities for reflection as they manifest themselves *at the time* of making a determinative judgment. The schema conceals this spontaneous art. Kant speaks of the schematism of pure understanding as "an art concealed in the depths of the human soul, whose real modes of activity nature is hardly likely ever to allow us to discover, and to have open to our gaze" (A141/B180–81). Yet, since the concepts of mathematics are themselves sensible, mathematics provides us at least with a sensible representation of that which cannot be imaged at all by the understanding. The schema is thus defined . . . defined as monogram (A142/B181), or a sketch that is always incomplete. It provides us with ciphers allowing us to apply the sensible to the intelligible, or vice versa, in relation to a given object of experience (i.e., what is necessary to think the roundness in a plate, or to intuit this roundness in a circle).[4] It aims at representing a reality of an entirely different order from itself, and succeeds in achieving *this* extravagance, because it contains this other reality in an embryonic form. The *next* extravagance is not allowed: there is no schema that would unite in some way an object of experience and an object of pure thought, hence a conflict of reason with itself arises when such a demand is formulated. By contrast, in the case of reflection, an object of pure thought is not even available in the first place, and so there can be no experiment at all. If there is anything like a principle of reflection, it must still be free to coexist with what this principle *is not at all*, since it cannot contain an incomplete sketch of what has to be realized. On balance, it seems therefore impossible to defend the legitimacy of the transcendental enterprise (i.e., the risk taken by experimenting with concepts) by reducing it to an exercise of imitation of the mathematical procedure. Rather, it is the mathematician engaged in the exercise of reflection, not the transcendental philosopher, who incurs the risk of transgressing the limits imposed by the realizability of possible objects of experience.

Yet there is every good reason to believe that this is the actual reason why the mathematician remains a model for the transcendental philosopher. Indeed, how does the mathematician move from the particular to the universal? Kant gives the following example (CJ 370):

> Suppose that someone coming to a seemingly uninhabited country perceived a geometric figure, say a regular hexagon, traced in the sand. As he reflected on this figure, working out a concept for it, reason would make him aware, even if obscurely, of the unity of the principle required for producing this concept. And so, following reason, he would not judge that such a figure is made possible by the sand, the adjoining sea, the wind, or even animals that leave footprints familiar to him, or by any other nonrational cause; for it would seem to him that coming across such a concept

> [a regular hexagon], one that is possible only in reason, is so infinitely contingent that there might as well be no natural law for it at all, and hence that such an effect could also not have been caused by anything in nature, which operates merely mechanically, but could have been caused only by the concept of such an object, a concept that only reason can provide and compare the object with. It would seem to him therefore that, although this effect [the figure] can be considered a purpose, it cannot be considered a natural purpose, but can be considered only a product of *art.*

Note that Kant's own exposition of the argument cannot quite escape the dilemma of "first reflection," since had he chosen a figure not in the sand, but say in the ice, a perfectly geometrical figure could have been compatible with a product of nature (a crystal). Kant acknowledges elsewhere that snow-figures could well be structured freely, that is, simply in accordance with chemical laws (CJ 349–50). But in this case the two concepts of art and nature are amalgamated, since we speak of the art of nature; in order to judge whether it is purposive, our own judgment must legislate as to whether it is *beautiful* or not, which is very different from considering how it is *produced.* In the present context, Kant immediately quotes his Latin and Greek sources: the man of the shore would exclaim: "*vestigium hominis video,*" "I see the trace of a man," a Roman account of the story of Aristippus, a Socratic philosopher, who reached the shore of Rhodes after having shipwrecked. In other words: we are still in ancient Greece, not in the icy mist of Königsberg; Kant's argument about the original reflection is most emphatically historical in nature.

The principle of the first reflection goes something like this. When someone ascribes the regular figure in the sand to a product of art rather than nature, he is forced to infringe the first testimony of his senses, namely, the impression that the country in which he found himself is uninhabited. And this is the enthusiasm that goes with any mathematical construction: it tempts us to find civilization even where any trace of its appearance happens to be absent. Such an enthusiasm immediately runs the risk of degenerating into extravagance. For according to Kant, the madman is he who persists stubbornly, against any evidence shared with others, in vindicating his viewpoint, since he then breaks off not only from experience, but also from the "subjective touchstone necessary for the exactness of our judgments in general." This touchstone consists "in founding my understanding upon that of another person, without standing aloof from the others, so that private representations will not serve a judgment which is somehow public."[5] As a breaking off from experience and communicability, the loss of what Kant calls *sensus communis* in the *Critique of Judgment* (CJ 293–96) is thus a loss of world and a loss of community. Still in the *Anthropology*, Kant writes: "He who persists in vindicating his personal opinion despite common sense, surrenders himself to a pure play of thought, where he sees, acts and judges not in a world experienced in common with others, but rather in a world which is proper to him (just like in a dream)." Thus no judgment, no science can be declared on its own supreme authority. Even the judgment of the land surveyor, which grounds mathematics in its originary basis, presupposes that there has been "beforehand a perception of universal coincidence with the judgment of those who devoted themselves to this task."[6]

What, then, is the transcendental account of the fact of civilization which leads from the primary factual community of men to the extravagance of merely postulated civilization? The extravagance cannot be a perfectly wanton act, because from the critical perspective, there can be no such thing as an abstract world of pure thought from which concepts are derived. The issue is to find out, and identify, no less than the necessary connection between our abstract thinking and the conditions under which we could think of a world in which we actually live. In contradicting the appearances of a desert country, the man of the shore happens to be not quite different from a first mathematician. Indeed, he participates in an extraordinary ruse of nature through which the latter achieves its final aim or ultimate purpose with respect to the human species, namely, culture—and more specifically the culture of *skill* (CJ 429–34). Skill is precisely that which inserts us in the world in which we actually live. How does nature achieve its purpose in us, even though this purpose cannot be willfully ours at the beginning of humanity since there is not as yet any principle of reflection to correspond to it? The answer is that the principle arises from a conflict which exists right *at the beginning* of humanity, distinct from an antinomy which *results* from the self-limiting process of thought.

Kant starts from what he considers to be a preeminent fact of early humanity: an originary difference or inequality among people. The majority simply takes care of the basic needs of life mechanically, serving the few who enjoy enough leisure to engage in science and art. But those who possess culture and keep the majority in a state of oppression are themselves unchained by enthusiasm, a state of mind which gives rise to insatiability from within. There comes a time when the culture of the upper class is overflowing, and gradually spreads to the lower class also, a phenomenon of cultural progress which, however, fosters further trouble among the two classes. Indeed, it stops neither insatiability among the upper class, nor the trouble resulting from violence from without among the lower class. This ultimate point in the diffusion of culture before the formation of civil society is "shining misery" (CJ 432). We have advanced from the unchained to the unruled, that is, the consciousness of that which is unchained has become unlivable. Now, while the lawful authority embodied in civil society may well alleviate the insatiability of the upper class by organizing public access to culture, the remaining violence from without has still to be countered. Humanity is thus in need of more than civil society: it needs a cosmopolitan whole, a system of all states, whereby the consciousness of what is required to live in accordance with art and science becomes *global*. Indeed, the free play of human will is only conducive to chaos as long as it remains attached to an individual; but no matter how deep the real causes of individual human actions may lie, a regular course (in the sense of historical sequence of events) may hopefully be discovered at the level of the whole species.[7] In regard to action, and provided the transition from the individual to the species is duly effected, concealment of ultimate causes does not get in the way of full development—unlike the schematism involved in cognition. According to Kant, it is not until such global consciousness comes to the surface that an individual's consciousness may earn the title of truly moral. Moral consciousness is the point at which the whole of humanity bespeaks its own cosmopolitan organization.

The Quarrel between Mathematics and Philosophy

This development from nature to culture determines both the foundations and the limits of speculative knowledge in its modern expression. Indeed, Kant's interpretation of the demands of human reason involves the existence of something like a ruse of reason that cooperates in some way with the ruse of nature. This forms the basis of the strange constructibility of the manifold from below upward, a construction that Kant feels compelled to articulate against Locke.

Whether our object is nature or whether it is freedom, thinking critically means incurring risks. In the case of nature, this is the risk of engaging in a first reflection without universal guide mark. In the case of freedom, thinking about freedom implies that our own thought must be assumed to be free: it has to assume the risks of its own autonomy, because it speaks on behalf of the destiny of the whole human species. But if the determination of this destiny may remain unrecognizable in the final analysis, what does the law of nature teach? Our directionless reflection will be provided with a minimum of guidelines, without which it could not even get started, if, contrary to the tradition, we try to answer about the question concerning the meaning of ends in the context of the law of nature, as distinct from the moral law.

We stand here at the threshold of the intellectual journey that led Kant to critical philosophy. In a work preceding the critical breakthrough by a few years, Kant could still absorb his later distinction between the lawgiver and the artist in a conception of the faculty of reason as "half inventive, poetizing and half inferential, syllogistic."[8] The critical conception of reason has accentuated the division between the two halves to the point of radical separation. But making full sense of the distinction, while retaining a common concern for ends for both types of law, seems at first to be rather devoid of significance. Does not Kant himself tell us that, by contrast with the purely intelligible world, "the sensible world holds out no promise that any such systematic unity of ends can arise from the nature of things" (A814/B842)? Furthermore, are we not today deeply imbued with Nietzsche's forceful observations (particularly in sections 6 and 11 of *Beyond Good and Evil*) that in Kant's philosophy moral aim is frankly the insidious motive underlying the theory of knowledge? On this account, perhaps the whole of the *Critique of Pure Reason* would be something like a cleverly cooked-up montage designed to clear the ground for practical reason's grand entry in the third cosmological antinomy. Such an interpretation is all the more tempting that Kant's first *Critique* cannot be taken literally as a theory designed to found modern natural science or constitute a theory of such science. Kant describes the Copernican Revolution in speculative philosophy as humans's attempt to "constrain nature to give answer to questions of reason's own determining" (Bxiii), which does *not* reflect the revolution undertaken by Galileo in natural science unilaterally. Galileo wrote it pithily enough: "nature has no obligation to men, it has passed no agreement with them," so that "the force of human authority on the effects of nature is nil, for nature is deaf and unyielding to our vain desires."[9] Similarly, Kant's distinction of the philosopher as lawgiver from the scientist as artisan reverses completely Galileo's notorious statement that "science is written in this grand book, the universe, which stands continually open to our gaze . . . [and] which is written in the language of mathematics."[10] For when Galileo tried to extricate himself from the scholastic concept of science, everyone knew and admitted that astronomy was already a branch of applied mathematics, just like optics, geography, or navigation for that matter.[11] Galileo's true motive was to reverse a judgment that had become commonplace in his days, namely, the neglect of mathematics in physical science by virtue of its having a dubious odor of practicality—the skill of an artisan. The reform of astronomy that he wants to undertake does nothing to elevate this practicality to any higher dignity. On the contrary, he charges the complex system of epicycles and eccentrics of pre-Copernican astronomy with having represented the world as a mere system of fables and fictions: "Nature takes no delight in poetry, . . . [for] falsehood is so abhorrent to nature that it is as absent there as darkness is in light."[12]

What, then, is an ideal philosopher of modern times, who does not simply live by the nostalgia of times past? The difference between nature and freedom hinges on the models that they offer. What distinguishes speculative philosophy, as it concerns itself with the essential ends of human beings, from mathematics and the natural sciences, cannot be understood without looking both downstream and upstream from the set of a priori knowledge furnished by human reason. (1) Looking downstream: we find mathematics, the concepts of which, Kant says, are "arbitrarily invented" (A729/B757) since they were given neither by the nature of my own understanding nor by experience. When it sets about to construct a figure, the imagination is directed by nothing determinate: the whole of mathematics is made up of "arbitrary syntheses." (2) Looking upstream: we find the fact that the highest achievements of speculative philosophy "cannot advance further than is possible under the guidance which nature has bestowed even upon the most ordinary understanding" (A831/B859). This highest point is what Kant calls moral faith. In his properly ethical work, the *Foundations of the Metaphysics of Morals*, he will go on to claim that philosophy has nothing to invent, nothing to add to popular consciousness. Even the properly speculative conflicts related to the nature of those principles that should account for the wisdom and honesty of men pale before that which we ought to do.[13] Philosophy can only attempt to elucidate, purify, and establish the a priori conditions of a fundamental perfection always available in ordinary life. What the popular consciousness pledges is that nothing can be held to be unrestrictedly good except goodwill.

Likewise in the theory of knowledge, Kant articulates the two heterogeneous elements of any possible knowledge (intuition and thought) in such a way that the perfection of the former serves as a model for the workings of the latter. But here the model can no longer be that of an originary familiarity. Kant distinguishes intuition from thought by claiming that whereas the former (as sensible) rests on affections, the latter (as intellectual) rests on functions. "Function" means "the unity of the act of bringing various representations under one common representation" (A68/B93). But as forms, the concepts that exert this function are *empty*. This is to be contrasted with the preconditions needed to be affected by anything: these are space and time defined as pure a priori intuitions, that is, as the sensible *content* which is pure because it necessarily filters the sensible impressions that reach us from the outside world. Each is given as essentially one and yet is represented as infinite magnitude.[14] There is an element of thought in these two forms of sensibility (and that is why they are both pure and a priori), but it is nonconceptual because they represent a pure originary content, and as such they contain an infinite number of representations *within*, not *under* themselves. Thinking by means of general concepts (the categories, with their necessary objective relation to the given in sensibility) emerges as *limitation* (B71) of the pure field of nonconceptual infinity.

Two fundamental consequences follow from this basic articulation. First, it is precisely because the space and time which make up the form of our sensibility are literally superhuman, by virtue of their infinity, that a bold Copernican revolution conforming the objects of experience to our faculty of intuition makes sense in the first place. The ensuing distinction between objects of experience and objects of pure thought corresponds to an inner division of human reason, a division forced upon it because such a split is the only possible way, for us finite human beings, to give expression to something as hybrid as nonconceptual thinking. Secondly, thought becomes conceptual only inasmuch as the modesty of the human condition can be retrieved, which is done by limiting the superhuman infinity of space and time as forms of sensibility. As a function, thought manifests itself as an expansion of its concepts, a striving toward retrieving the originary infinite perfection of pure intuition. But note that this striving is bound to fail to make contact with its model, if at least it is to be consistent with the immediacy of our own unavoidable sensibility. Thus, Kant articulates the transition from general to transcendental logic in such a way that what counts as infinite judgment in the former becomes the category of limitation in the latter (A71–72/B97; A80/B106). As it turns out, there is only one concept that has the capacity of rejoining without loss the forms of human sensibility: this is the concept of *ens realissimum*, or universal concept of a reality in general (transcendental ideal) which, Kant claims, *cannot* be divided a priori because it not only "comprehends all predicates *under itself*, it also contains them *within itself*" (A577/B605). This concept of the highest reality is of course the being of all beings, and viewed from this standpoint, limitations amount to negations (A578/B606).

This latter concept of transcendental ideal sends us to the highest point of the *Critique*, that is, God as the third of the three Ideas of Reason. As for the Idea of World, it deals with the very process of transgression whereby reason aims at making contact with the given infinity of nature in general. Heidegger contends that human reason according to Kant always means finitude; the Ideas of Reason thus also indicate the point where our own critique of the Heidegerrian interpretation of Kant may begin. Finitude, Heidegger argues, is not simply an accidental feature of pure reason, but instead, its finitude defines it as necessity to make itself finite.[15] It is true that the transgression of limits is the character of finitude. But we should not overlook the fact that this transgression is a regression toward an original condition which is formally infinite. Of space as a given infinite magnitude, Heidegger says that it is infinite because it is essentially (infinitely) different from its particular parts. As a unified whole, it is prior to its parts, which are limited in it.[16] Infinity is thus interpreted here in conformity with the priority of the ontological ground over ontic beings. Instead of a movement of return to the original infinite, which is proper to dialectical reason, Heidegger only acknowledges that pure intuition is an originary act of representation, allowing what is intuited in it to spring forth. In his view, it is essentially the primacy of intuition that indicates the finitude of human knowledge, "despite the reciprocity of the relationship between intuiting and thinking" that Kant seems to hammer in throughout the *Critique*.[17] On this account, failing to do justice to the formal infinity of originary intuition, Heidegger argues that, in Kant's doctrine, only divine knowing knows of no limit, precisely because it is pure intuiting. In fact, Heidegger conceives of originary intuition as creative, which places it immediately in the realm of divine intuition: he renders *ursprünglichen* by *entspringenlassenden*, a letting something spring forth.[18] This interpretation passes over the formal infinity of our own human intuition. The difficulty of this concept is formidable: how can an infinity be "giving," and not immediately creating, since we would expect something infinite to have everything both under and within itself?

Following Kant more strictly, we see that this invasion of infinity, as formal precondition for any cognition, is the counterpart in determinative judgment of the absence of first reflection in the realm of reflective judgment. In the act of self-positing its own condition as ultimate condition of any possible knowledge, human reason bears witness to an infinity which it cannot make intelligible. The critical problem is thus the following: as long as we remain concerned with objects of experience, how can our thinking them by means of limitation be other than, or not yet, absolute negativity? How does this thinking fulfill an effort which, when all is said and done, would look like nothing by comparison with the transcendental ideal? In bringing various representations under one that is common to them, our thinking does again what in pure sensibility has always and already been done. Thinking is judging, in the sense that a relation of a subject to a predicate is brought about (A6/B10), and therefore a judgment is nothing other than "the mediate knowledge of an object, that is, the representation of a representation of it" (A68/B93). Not surprisingly, then, the faculty which enables us to couple the sensible and the intelligible (i.e., imagination) works essentially by following the model of *repetition*: this is transcendental schematism.

Admittedly, mathematics furnishes the first example of such coupling. A mathematical proposition (such as $7 + 5 = 12$) is both synthetic and a priori, because in order to formulate such a judgment the rule of addition is given entirely a priori (as something merely possible), whereas the actual operation is synthetic because it takes time—namely, time as an inner sense in which the mathematician constructs his concepts in determinate intuitions (B15–17). But in the transcendental schematism, only a striving toward this complete symbiosis of the synthetic and the a priori is allowed. The schema is an operation whereby the concept becomes the rule of the object (A140/B179), that is, the operation itself becomes no more than a rule: it is, as it were, *twice* a priori because it can no longer aim at any intuition in particular; it represents the particular as general (or the "universal procedure of imagination in providing an image for a concept"), and as such it *cannot* be an operation in the mathematical sense of synthesis. The special intuition (or image) is distinguished *intuitively* from the schema (or rule for the production of that image) by the fact that the latter is *much larger* than the former; when the connection between the concept and its image becomes loose because the image is very large, the schema takes over inasmuch as it provides the method for the representation of an image that remains in conformity with the concept. On the other hand, where the spatial image grows larger, its counterpart in *time* grows thinner. Indeed, the time required by the schematic operation is no longer the formally infinite time of inner sense, but rather the more strictly "human" (finite) time which is always and already implicated in the stream of experience. For example, "the schema of magnitude is the generation (synthesis) of time itself in the successive apprehension of an object" (A145/B184); the other time determinations are the filling of time and the connecting at all times. How do we go from time as the infinite form of intuition to the time which is closer to our actual life?

If we apply the doctrine of schematism retrospectively to *mathematical* concepts, we will have a transcendental interpretation of mathematics that differs from the actual operations performed in mathematics. These operations are always adequate to the concepts that make them possible; when I do mathematics, I do not comprehend but I simply apprehend magnitudes, because these magnitudes are always presented as relative, by comparison with other magnitudes of the same species. In this way, as Kant puts it in the analytic of the sublime of the *Critique of Judgment*, apprehension is not problematic from the transcendental viewpoint, because it can be carried out to infinity without ever coming up against a maximum that could not be overtaken (CJ 251–52). On the other hand, from the transcendental viewpoint, "no image could ever be adequate to the concept of a triangle in general" (A141/B180), that is, the triangle in general could never be completely comprehended (collected together and held together) in one image. Yet the rule for the production of such an image overlaps perfectly with the concept, because the figures dealt with in mathematics are still what Kant refers to as *pure* figures in space. Thus, in mathematics, the schemata that correspond to certain concepts are virtually indistinguishable from the concepts themseves.

However, for the *empirical* concept (such as the concept of dog), comprehension cannot be simply absorbed in the process of apprehension. The schema of a sensible concept is now only a sketch in the sense that, as Kant says, the rule is only "compatible" with the concept. What does compatibility mean? In the example of dog, it means that the dog does have a determinate *Gestalt* (it is a form, though not a pure one), but given the structure of our thinking, this determinateness can only be the sketch of the complete dog. The dog becomes ultimately identifiable as dog, because the sketches that we form on the occasion of the perception of a dog are pretty much similar at all times: due to this similarity, we can refer to the sketches as rules. Indeed, there is nothing vague in a sketch, since Kant emphasizes that it leaves a definite trace (*verzeichnen*) in intuition. To be sure, the contour of a thing as presented in intuition may be insufficient, and therefore "miss the mark," in that it may not enable us to identify the thing that stands before us as that very thing—we identify the dog no better than as a four-footed animal. But this is nothing other than the character of an object of experience: *our* deficiency only reflects a deficiency in the *thing itself*, which is always given (prior to our determining it) as an index of other more or less similar things. That is, our communication with the natural world outside (through sensible concepts) is the coming into being of an encounter between two negativities. We have an empirical concept of dog, but since we cannot proceed to a transcendental deduction of an empirical concept, it is also true on Kant's terms that we *never quite* have it.

The existence of this residual gap is the essential ingredient of transcendental reflection. For in the case of the concepts of understanding that are themselves pure (i.e., the categories), any image corresponding to them has become so large that their schemata "can never be brought into any image whatsoever" (A142/B181). Their spatial support having vanished in virtue of being so large, only the temporal counterpart to an image remains—what Kant calls the transcendental product of imagination as *Zeitbestimmung*. What is the connection between the absolutely complete time that inner sense is and this fragment of time? If the identification (or a priori conceptualization) of the object by means of schematism is to comply with the general character of conceptual thought as limitation of a non-conceptual thought, then a schema as *Zeitbestimmung* means the categorial constitution of an object *in*, rather than *of*, time. The distinction is crucial, because it touches perhaps on what is ultimately finite in the human condition: in contrast to any putative afterlife in the heavens, what characterizes our own lives is precisely that the constant potentiality for change—in particular, any act of thought involving a moment of reflection—takes place *in* time.[19] Inner sense is not, as it were, "manipulated" by the category as it is applied to the schema. Rather, inner sense retains its infinite untouchability, because the time which is made categorial through the process of schematization can only unfold what in inner sense is always and already closed upon itself within its infinity. Mathematics provides a measure by which the failure of any infinite synthesis via schematism is secured in advance. An infinite synthesis would occur in the case of a *magnitude* (not a manifold) being given as infinite in space and time. By virtue of this infinity, we would expect the distinction made concerning the time determination between "in" and "of" inner sense to collapse. But Kant argues that "an infinite whole . . . does not represent *how great* it is, and consequently is not the concept of a *maximum*" (A430–32/B458–60), because it can only be represented as the addition of a unit which always exceeds any given number, however large. Any attempt to fix the absolute magnitude would remain dependent on the chosen unit, which of course is arbitrary. In other words, the arbitrariness of any mathematical synthesis owes its very arbitrariness to the fact that, when mathematized, the infinite space and the infinite time of experience lose their fixed givenness. Infinity as a given magnitude dissolves completely; all that remains is that "the successive synthesis of units required for the enumeration of a quantum can never be completed."

The originality of Kant's theory comes into relief when compared with that of the modern champion of mathematics taken as the model of all philosophical understanding: Descartes. In the third of his *Rules for the Direction of Mind*, Descartes argues that knowledge in general (whether mathematical or other) can be obtained in no other way than by clear and evident intuition or by certain deduction. This claim might give the impression that deduction is a conceptual rather than intuitive operation, for example, the deduction of theorems from a set of axioms, or conclusions from premises by logical inference. Yet, this is not quite the case, since Descartes' attack against the traditional syllogism leads him to argue (in the eleventh rule) that a simple deduction of one fact from another is performed by means of intuition. How, then, are the two operations of deduction and intuition to be distinguished from one another? Intuition, Descartes goes on to say, is the instantaneous apprehension of a single simple nature, such as a number or geometrical magnitude. As such, however, intuition is insufficient to establish continuity or relation between the objects which have been clearly and distinctly intuited by the light of the mind. The continuity (of, say, a series such as A:B::B:C) is secured by deduction: the mind moving from one intuited simple nature to the next while holding together in memory the succession of instantaneous intuitions as a whole. Whereas intuition is the instantaneous apprehension of a simple nature, deduction is the successive movement of the mind by which simple natures are held together in the memory. But memory, Descartes warns us, is weak: it is only "in a sense," as he writes in the third rule, that "deduction gets its certainty from memory." This dubiety to which the weakness of memory exposes deduction is remedied by what Descartes calls a continuous movement of thought (rule 7). In the case of long series of magnitudes, the knowledge of the relation between the first and the last requires that "I shall run through them [each magnitude in its relation to the next] several times in a continuous movement of the imagination, simultaneously intuiting one relation and passing on to the next, until I have learnt to pass from the first to the last so swiftly that memory is left with practically no role to play, and I seem to intuit the whole thing at once." Thus, by such continuous movement, the successiveness of deduction approaches the instantaneity of intuition as a limit case, enabling the role played by memory in deduction to approach zero. Because deduction may approach intuition as a limit case, Descartes characterizes it as "a completed process . . . [which] no longer signifies a movement but rather the completion of a movement" (rule 11). As a completed movement, deduction is essentially intuition. Just like Kant's, Descartes' rule of imagination "misses the mark," but only because it is an expedient (something practical) that can be dispensed with under certain conditions. Kant's rule is transcendental precisely inasmuch as the practical sense of rule in mathematics cannot function as the model of all knowledge; the completion of the sensible manifold by an act of thought is not possible. Thinking does not allow any short cut, or does it any invite any.

Instead of furnishing a prototype of all schematism, mathematical construction as envisaged by Kant does no more than partake of the imperfection of all material copies found in the world of experience. Only the original possibility of that which in itself includes all reality (the transcendental ideal) is entitled to lay claim to the status of archetype, and mathematics enjoys no special privilege in making us any closer to it. Just like all real things in their manifoldness, "all figures are only possible as so many different modes of limiting infinite space" (A578/B606).[20] Hence the distinction concerning magnitudes in general that Kant makes between the pure image and the pure schema (A142/B182). Time *and* space are called pure images of all magnitudes, whether for inner sense in the case of time or whether for outer sense in the case of space. This is simply a consequence of the fact that mathematics deals primarily with pure figures in space, since even counting presupposes at the very least the representation of points in space for each number; when large numbers are dealt with, the corresponding images are no more than convenient signs in which the original relation to the magnitude is wholly arbitrary—it does not have the necessity of a schema. But the pure schema of magnitude is not a determination of these pure images, because it results from "my generating time itself [i.e. neither the image nor the pure image] in the apprehension of the intuition" (A143/B182).[21]

Whence the vindication, first inaugurated by Plato and then rejuvenated at the dawn of modern philosophy, for viewing mathematical harmony at the basis of the mechanism that sustains the ultimate root of all things? Human thought has the strange natural inclination to hypostatize (elevate an object of thought to the status of real object) even in the case of the highest of its ideas, namely, the transcendental ideal, as if it could not rest satisfied with the irrealizable idea of a maximum (A580/B608). The concept of God is a direct expression of this illusion of reason, which is based on a natural, not on a merely arbitrary idea. Could it be the case that mathematics fell prey to a similar process?

It has become customary to characterize the nature of modern mathematico-physical science by means of a radical shift from the ancient view of mathematics.[22] This shift alone would explain how it became possible to change the main object of measurement in modern natural science: while the ancients focused on fixed spatial proportions, the moderns turned their attention to the phenomenon of motion itself. Greek geometry had considered relations of unchanging figures and bodies, but the moderns found that it is possible to represent the geometrical form itself as a function of variables, that is, as a phase in their continuous growth, and so to formulate the laws of its "generation." The true objects of mathematical cognition are now these generative, determining laws, instead of the descriptive or determinate forms of the Greeks, which have lost their independent status for that of transitional limits. The "forms" of being are no longer what the Greeks viewed as terminal products, but those of the continuous processes of nature. Process is now defined by its own form and in no way by its end.

This is how the modern philosophers of nature speak about the Greeks in order to distinguish their own achievement. But did not the moderns overrate their differences with the Greeks precisely in order to point them out with more punch? Kant refers to the ancient geometers who were eager to investigate the properties of such lines as conic sections, "not letting themselves be disconcerted if asked by narrow minds of what use such knowledge might be" (CJ 363). Such figures allow us to solve a multitude of problems, because they are fertile in many infinitely splendid properties which can be generated spontaneously, as it were. The Greek geometers lived in the happy innocence of origins, because the actual use of such spontaneous constructions has been revealed quite spectacularly in modern times, when mathematical physicists were finally able to show that the law of terrestrial gravitation allows us to represent the trajectory of heavy bodies by means of the parabola, which belongs to the class of such curves; similarly for the investigation of the ellipse, which was undertaken by the Greeks without the slightest consideration for its later application to the gravity of celestial bodies. Thus, the intellectual purposiveness proper to mathematics does not have to be regarded as being based on a definite purpose; the possibility of a mathematical concept owes nothing to its own purposive use or application. In fact, if a schema happened to be immediately available separately from the concept, the variety of heterogeneous rules used by the mathematician would immediately point toward a unity recognizable outside us and distinct from our presentational power (CJ 364). But mathematics has the peculiarity of making progress through this very anarchic absence of unity among rules. Through reflection I am only given a principle of unification, whose truth I may well be cognizant of a priori, yet the special foundation of it is *not* external to my concept or representation. Hence the perpetual amazement (*Verwunderung*) that sustains and underlies the progress of mathematics, which is a shock exclusively in the mind resulting from the unexpected extraction of all sorts of cognitive uses of geometric figures from their simple construction. The shock gives us an inkling of there being something else above and beyond any possible sensible representation of ours, a something which we ignore and which, however, might well conceal the ultimate basis for the harmony between our reflective principle and the nature of things.

Thus, in modern times, we attend to the successful application of ancient mathematical ideas, by which a schematization is, as it were, *added* to concepts which originally were quite alien to it. Kant speaks of the amplification or expansion of the mind obtained by such continuously rejuvenated amazement (which he calls admiration) that has the force to transcend the doubt generated by the clash between our reflection and the way things are. Clearly enough, Cartesianism has managed to give an ultimate expression to this renewed capacity found in mathematics to overcome the most resilient doubts about the reality of our constructions, because in this doctrine the way things are is supposed to be an unmediated outgrowth of our own mathematical constructions. But precisely in order to dispel the temptation of equating the amplification of the mind with any kind of encroachment on the being of things, Kant urges us at this point to do away with one of our deeply ingrained prejudices, which consists in calling beautiful those properties of geometric figures and numbers which display such a priori purposiveness. He writes: "For example, people will speak of this or that *beautiful* property of the circle, discovered in one way or another" (CJ 366). But it is not by aesthetic judging (a judgment without concept indicating the free play of our cognitive faculties) that we find such a property purposive. Rather, the judgment is intellectual in accordance with given concepts suitable for an infinite diversity of purposes. That is why we should rather speak of "the figure's *relative perfection*." What does relative perfection mean? Kant has argued in the section of the *Critique of Judgment* concerning the ideal of beauty, that the only thing in the world that admits of the ideal of perfection is man himself, precisely inasmuch as the whole of humanity may be present in his person (CJ 233). On the other hand, the mathematical estimation of magnitude never exhibits more than relative magnitude, whereas an aesthetic one exhibits absolute magnitude to the extent that the mind can take it in one intuition—that is, only in the case of the aesthetic estimation of magnitude can we say that there *is* indeed a maximum (CJ 251). Thus, mathematics in modern times has bestowed upon the mind an amplification such that mathematical perfection could be supposedly justified as absolute. This is the fanaticism of modern times: the mathematician has now monopolized attention to the extent of being regarded as the "ideal philosopher," exemplifying through the realizability of his concepts the true vocation of the human condition. It is, however, a case of usurpation on humanity's rights and each individual's claim to autonomy. For no matter how free the employment of our faculties (cognitive or other) may be, we can do no more "than draw on the same sources from which our predecessor himself drew" (CJ 283), failing which the process whereby our concepts become possible would be transformed into a mechanism of mere imitation. We moderns have recognized the definite purpose which underlay the mathematical constructions of the ancients, but the price to pay for this must be an explicit acknowledgment of a chasm separating being from knowing; the ancients could afford to ignore it only because the true goal of their constructions remained hidden to them.

To be sure, when Kant undertakes to justify his own Copernican revolution in metaphysics, he speaks about the examples of mathematics and physics as being sufficiently remarkable to suggest an imitation of their procedure, so far as an analogy may permit (Bxv-xvi). But what did a fictional first mathematician do in the earliest times of humanity? What is the method which came to his mind? Basing his argument on geometrical figures, Kant begins by rejecting two possible accounts (Bxi–xii). (1) Mathematics did not originate from the sheer manipulation of arbitrary pure concepts. Take the defining property of the isosceles triangle—a triangle in which two of the sides are of equal length. The possibility arises of demonstrating this property by reasoning from some of the consequences of this equality in length for the equality in angles; for instance, that the two angles at the base are equal because the two sides from the vertex are equal. A demonstration for these properties, Kant tells us, could not be discovered by just toying, as it were, with already existing concepts of equality (equality between angles, equality between lengths). Rather, concepts of this kind had to result from the very discovery of the properties by means of demonstration. The isosceles triangle was not traced in order to comply with bare concepts that we already possessed, as if the invention of mathematics would proceed from the mere decision to represent in a figure a certain number of preexisting concepts. (2) Nor did mathematics originate from an initial attempt to discern the properties of some given figure. That is, it would be wrong to believe that the figure had always been there, as something to be inspected, and that nothing more was required but that someone would come along one day and tell us that such and such are properties of this triangle. Of course, we see around us a certain number of shapes which come very close to the perfect isosceles triangle, but Kant argues that the necessity underlying mathematical reasoning as such can have nothing to do with the attempt to perfect what is already around us. Rather, neither the concept nor the figure were given in advance. But the way mathematics emerged teaches us something crucial about any possible relation of human reason (as containing concepts) to experience (as displaying figures). Some necessity must be shared by both reason and the content of experience. If the concept (of, say, equality between lengths) is supposed to be given in advance, then there is no way of seeing what is necessary in experience itself for the concept to be realized in it. If the figure is supposed to be given in advance, then what about the mind which imposes some necessity (a rule of reason) on it (e.g., the two angles at the base are equal because such and such)? Both views are inadequate because they mask the actual nature of our concepts. Mathematics became a science, that is, it forced universal agreement among rational thinkers, when and only when the necessity which accompanies mathematical concepts could emerge as a consequence of the fact that we recognize our own construction as ours, as a product of human reason. The figure was not given, but constructed precisely in such a way that the construction corresponded to what had been put in it by the mathematician himself; and what had been put in it in this way was nothing other than the concepts needed for the construction (e.g., equality between lengths, etc.).

The difficulty with Kant's argument is that we are always tempted to consider our concepts as preexisting, precisely because they belong to reason alone. But according to him these concepts were *nothing* until the first construction of some figure in *experience* was made. In any mathematical presentation, the coming to presence of concepts is not prepared by anything that contains them potentially in some way; the presentation forces our assent (i.e., mathematics always strikes us as *the* paradigm of absolute universality) because of its peculiarity of making present an original absence; an original concealment comes to the surface without the presentification being enough to change anything in the original concealment. In doing mathematics, it is not just my mind, but the mind of the community of rational thinkers which gets immediately amplified because what has been added is an empty generality. Thanks to this interplay of presence and absence which is not mediated by anything, Kant brings out an essential character of the primary concepts of any mathematics (such as axioms or postulates), namely, that their necessity imposes itself on our apprehension without it being possible to reduce it to something else. But at the same time, the lofty significance of this inner necessity for the mind emerges from its being rooted in an historical fact of utmost fragility. Mathematics came into being at the end of a chain of circumstances that might very well never have taken place; it depended upon "the happy thought of a single man," whether Thales or any other thinker, that "a new light flashed upon [his] mind." That is, the first geometers had not taken the trouble to write down and keep a record of each step in the reasonings that led them to their first theoretical achievements. The success of the first mathematical demonstration is a correlative of the fact that the first mathematicians did not bother to reconstruct the steps of their demonstration: they did not because they could not, owing to the impossibility of expressing anything with regard to the prior conditions of possibility.

Mathematics is an inner exercise of the mind left to its own representations. As such, the *whole* person of the mathematician must have been engaged in this initial project, which is precisely what surrounds it with a higher moral glory. For this engagement of the whole person makes it impossible that it could ever be taken away from him, just like moral disposition (*moralische Gesinnung*), which Kant believes can never be lost since it cannot be communicated at all (A829/B857). Of such disposition, Kant says elsewhere that it is innate constitution,[23] which does not mean that human being has not acquired it or that he is not the originator of it, but rather that he has not acquired it in some determinable time. Through this disposition, I can only say: "I am morally certain that . . . ," instead of: "It is morally certain that . . . " Thus the first mathematicians did not take the pain to record their steps for the sole reason that they were responding to the call of their own inner disposition as legatee of humankind. Schematizability or constructibility was added later as an ideal of mathematical intelligibility, in an attempt to make communicable what is in fact radically noncommunicable (because always and already evident). This tension alone explains the fundamentally arbitrary character of mathematics, since constructibility may well be no more than an act of persuasion, in the sense of displaying an evidence which is already accepted. Indeed, persuasion is a holding of a thing to be true (*das Fürwahrhalten*) which has its ground in some special character of the subject rather than in objective grounds valid for everyone (which would be conviction) (A820/B848). Kant argues most emphatically that it is only a matter of presumption "that the ground of the agreement of all judgments with each other . . . rests upon . . . the object," since "the touchstone whereby we decide whether our holding a thing to be true is conviction or mere persuasion is . . . external, namely, the possibility of communicating it."

If the fanaticism of mathematical perfection has thus to be lowered to the level of merely relative perfection, what is it that is radically nonconstructive in transcendental knowledge? How exactly does it bestow upon the mind the amplification which will actually fill it, instead of usurping the essential ends of human reason? Whereas in the case of mathematics, only the "memory of the revolution" was enough for its first constructions "to survive the tide of oblivion," metaphysics finds itself in the uncomfortable position of having constantly to redefine the direction in which to go: "ever and again we have to retrace our steps" (Bxi; Bxiv). Kant defines axioms in mathematics in such a way that the spontaneity of the first construction remains present: as synthetic and a priori principles, they are "immediately certain" (A732–33/B760–61). That is, when using the axioms for some construction, I do not know in advance what I intend to think (and have no determined purpose): constructing the concept in intuition means that I construct from above what is given in intuition. By contrast, in order to find a principle of pure understanding that will be absolutely first, in the sense of a principle of principles ("Principium dieser [mathematischen] Grundsätze," A160/B199), a change of *direction* has to be attended to—instead of proceeding from intuition to concept, we now proceed from concept to intuition, that is, we construct the manifold from below upward. In transcendental philosophy, the deepening of the sense of mathematics leads to reversing the direction of its natural reflection. As a result, we cannot afford the luxury of combining concepts while passing over a mediation, a third something, which *cannot* be immediate, namely, a *determination* which will make our cognitive judgements truly determinative. It is a *time* determination, spontaneity is lost. Kant writes explicitly that only mathematical axioms enable us to combine representations "both a priori and immediately." Now, at the very beginning of the *Critique of Pure Reason*, Kant points out that it is only "in the order of time" (B1) of ordinary experience that it can be said that no knowledge precedes this experience. The reflection on that part of knowledge which does not necessarily arise out of experience is bound to upset the natural order of time, through which we have no knowledge antecedent to experience. There are essentially two ways of disturbing this natural order—speeding it up or slowing it down. That is, there must be something quite remarkable about the present of any appearance that allows it to be stretched or shrunk so that it can be connected in some way with what lies under it—that which in the manifold has to be unfolded. Spelling out this feature is the main gist of the first two (Kant calls them "mathematical") principles of pure understanding.

In the first of these principles ("Axioms of Intuition"), Kant begins by making a primordial distinction between the mathematical and the transcendental sense of an axiom (A163–70/B203–12). The distinctive mark of the latter lies in the fact that the genesis of intuition is never wiped out, that is, its continuity is always transparent under the result (an extensive magnitude) following my production of an intuition. For instance, in order for a line to arise as an outer appearance that can be represented in sensible intuition, I must first draw it in thought. Thus, the production of intuition for a given extensive magnitude leaves a trace behind it, open to inspection at any time. The presence of a trace indicates that this production is no creation *ex nihilo*; a distance installs itself between the subject and this object of knowledge, since I must be able to relate to my own production as an *outer appearance*. But now we want to establish the transcendental validity of the principle of that genesis, and we find that those magnitudes which correspond to it leave no trace behind them. These are the intensive magnitudes, as dealt with in the "Anticipations of Perceptions." They stand between *two* beginnings, namely, the instant at which an appearance begins to be perceived is preceded by another instant at which the appearance must have begun for its own sake, without it being possible for me to be aware of all the infinitely small degrees that, starting from zero, led to the beginning of my actual perception. Any real (as intensive magnitude) presents itself to me in such a way that it has always and already speeded up in order to approach me; if I could slow down the process, then—and only then—would the infinitely small degrees below the threshold of actual sensation become perceptible.

More generally, the two types of magnitudes, extensive and intensive, can also be distinguished by reference to addition or subtraction. Magnitudes apprehended by us as extensive are generated by the addition of parts one after another; this, in fact, is the manner by which an empirical intuition is produced, pure intuition providing the arena through which the addition is possible. By contrast, apprehension by means of mere sensation takes place in an instant and not through successive synthesis. That is, sensation is prior to the forms of sensibility themselves (space and time): it does not proceed from the parts to the whole, but is rather given in its totality in that instant which can no longer be one among other instants of pure intuition. This magnitude which corresponds to sensation, since it is prior to the deployment of the succession of time and the coexistence of the parts of space, can be apprehended only as a degree; it is a "flowing" magnitude, because it "occupies" or "fills" time, not by the series of numbers which can always be traced back, but by progressing constantly and slipping away without leaving any trace behind. Whereas the transcendental principle of extensive magnitudes "greatly enlarges our a priori knowledge," the anticipation of perception is the principle of "an intensive magnitude which can always be diminished."

In order to make sense of the nature of human reason according to Kant, we had to look both downstream and upstream from the set of a priori knowledge furnished by human reason. Now, there is one aspect of being which shows itself as it is, that nothing in the freedom of knowledge can absorb: the empirical concept of a moving body in space. Kant investigates the implications of transcendental knowledge for the intelligibility of scientific knowledge in its own right in the *Metaphysical Foundations of Natural Science*. Since the main character of modern science lies in the way Galileo and Newton have used mathematics as a tool to unlock the secrets of the physical world (in particular, the nature of motion), Kant's task in this work is really extraordinary since he has to combine the inner degradation of sense diagnosed in the case of mathematics with his own transcendental movement elevating thought toward ultimate ends. Is, then, the theory of motion a point at which the two opposite streams of knowledge are reconciled? Yes and no. Yes, because in order to be able to construct a concept, the condition of presentation cannot be borrowed from experience; but motion can only be given to us in experience, it cannot be deduced transcendentally at all.[24] No, because the being which shows itself as it is, is motion in *space*, not in time. This double condition of natural science is brought to light by reflecting upon the limits of constructibility in reference to the theory of motion.

Kant argues that an absolute transparency of physics and mathematics is available in one case only, that of phoronomy (or kinematics)—the science of the velocity and direction of purely quantitative moving bodies, in which physical bodies are arbitrarily emptied of all qualities that are not directly geometrical. On this account, the principle of the composition of two motions into a third is effected easily enough by means of geometrical construction (the familiar parallelogram construction). But in the case of dynamics, the qualitative resistance of one body to being moved by another is taken into consideration. Therefore, a distinction is needed between spatial extension, which for a body consists in occupying a space (*Ausdehnung*), from the filling of space (*Erfüllung*).[25] In order for any synthesis involving the filling of space to be possible, a specific force has to be ascribed to the body, a force which could never be fully constructed (at least by us). Indeed, it is given to empirical consciousness as a degree, and since every degree is expressed by means of a number, abstraction is made of its continuous genesis, contrary to the straight line of geometry (or the kinematic diagram of motions) in which the continuity of the genesis is still there, transparent under the result.[26] A totally mathematical physics is thus a case of abused imagination, even if within its own limitation it can present itself as a completed idealization. This is the Cartesian dream, in which motion is a mode of figure, a dream which Newton has crushed by showing that no body, in gradually changing its form, could possibly be transformed into the species of bodies that gravitate;[27] thus, that from the sole geometrical form, the diversity of phenomena (together with a representation of their physical cause) can never be derived.

Kant arrives at two fundamental forces constitutive of any filling of space, namely, attraction and repulsion. He claims that, precisely by virtue of the limits of constructibility, the intelligibility of these originary forces in accordance with their a priori possibility lies beyond the horizon of human reason.[28] The limits of constructibility are plainly evident in the fact that the process by which we discover and differentiate between attraction and repulsion *cannot be confused* with the nature of the objects themselves so apprehended. Through repulsion a body is impenetrable, which gives us immediately a determined concept of the object in space since its shape and size fall within the range of our sense of touch. Not so in the case of attraction, which is apprehended through the sense of sight, and so cannot reveal to us a matter of determinate volume and shape since this force always emanates from a center that matter masks as it falls towards it.[29] Touch and sight are separated from one another just as the movement of approach cannot be deduced from the movement of withdrawal; no arbitrary exchange (by means of a construction yet to be found) prevails between resisting the withdrawal (attraction) and resisting the approach (repulsion). When constructibility is transposed from the phoronomy to the dynamics, it thus becomes a constraint which blinds us to the first principles.[30] Since this constraint is unavoidable in mathematical physics, we are led to the conclusion that, over and above the difference in apprehension, repulsion and attraction can only be identified are two originary forces. To an absolute difference between two senses there corresponds an absolute difference between two forces. A transcendental synthesis is not available to bridge the two differences. It is transcendentally impossible to reduce the natural world to anything more fundamental than two originally heterogeneous forces.

The movement of thought within the pair of transcendental principles of pure understanding, which make up the "mathematical principles" of any rational thought, has thus opened up a speculative field of reflection about *origin* and *totality* which is completely irreducible to natural science in particular, and to nontranscendental critique generally. In the remainder of this book, we shall examine the nature and implications of these two concepts within their own horizons, that is, the special kind of knowledge that they bring about. In the "Anticipations of Perception," Kant shows that, in the present of a perception, the very presence of an object is loaded with an (unconscious) sphere of absence. However, as we proceed to enlarge the sphere of the perceptible to the trans-empirical (the not immediately given), and move on to what Kant calls "dynamical principles" of pure understanding, we shall attend to a deployment in time of the sphere of presence and absence. Ultimately, a separate sphere of absence in the remotest past will define the transcendental concept of unconditioned.

Of Shining Misery in Modern Times

Why should transcendental philosophy be preoccupied with trans-empirical experience? In order to understand this, we must now look downstream and upstream from Kant's philosophical system itself as a historical achievement. The transcendental turn stands between two radical beginnings that touch on the whole project of modern philosophy, both in its inception—Descartes' cogito—and in the radical reenactment thereof undertaken by Husserl's phenomenology. In this way, Kant's philosophy captures the insecurity proper to all authentic beginnings.

The Cartesian cogito is prior to the knowledge of the universe; and furthermore, in order to accomplish itself, it cannot be preceded by any such knowledge. The knowing subject discovers itself in itself by willfully abrogating any relation to the world. It then takes possession of its essence, at a mental time and place of perfect plenitude, filled with thinking substance. Descartes is therefore unable to ask the questions "Where am I?," or "What is the position of my essence with regard to other essences in the cosmos?," in any meaningful way. My position in the cosmos is not a floating point, because my thinking substance is self-sufficient, it is not relative to any other; the Cartesian cogito is not a cosmic concept. Nothing is more different from this than Pascal's acknowledgment that the starting point of knowledge is the knowledge of our own miserable position in the universe. The first awakening to the question of sense enhances the relativity of my position: "I am seized by terror like a man who has been carried, while asleep, to a deserted and frightful island, and who wakes up not knowing where he is, and without any means of getting away from there."[31] In order to know who I am, I must know where I am. Consequently, I am compelled to relate myself to the totality of things. I must embrace in thought all there is, as a preparation for the encounter with the very being which is the reason of what exists. But Pascal knows that this possibility is not quite a real one; we do not really comprehend more than our misery. When it leaves its island, the place of thought shrinks as the sphere of its contemplation grows. Ultimately all images are blurred, whether they originate from the eye or from the spirit. The experience of self-knowledge through the experience of the universe ends in total failure.

In direct contrast with Kant, and as a response to the urgency of the crisis of contemporary times, Husserl, in the *Crisis of the European Sciences*, develops a notion of transcendental subjectivity aimed at overcoming the Kantian foundation of knowledge which, according to him, would do no more than ratify the infinite progress of modern science toward being. This notion should allow us to reconstitute our lost intimacy with being by exacerbating the power of the ego. The point is to retrieve a natural sense of hierarchy (the classical tree of knowledge) in a world which would not be permeated by the open infinity of science, even though this world of ours has irremediably lost every trace of the ancient cosmic order. The world will thus remain as a unique world-horizon deprived of an original plurality of sense, whereas the overflowing ego presents itself as large enough so as to function as a "kingdom of something subjective which is completely closed off within itself."[32] Husserl is perfectly aware of the risk. There is a sort of choice to be made at the outset between two possible propositions to serve as candidates for the ultimate principle of all grounding: on the one hand, "the world is," as containing the manifold and ever renewable multiplicity of world experiences, and on the other hand, "I am," as touchstone of all genuine validities.[33] The former *seems* to include the latter, but Husserl inverts the relation and goes on to advocate the reduction of the natural attitude (the transcendental *epochē)* by virtue of the incomplete character of the world. Indeed, the perception of a thing in the world only gives us a series of profiles; at the very most, the thing before us is an Idea of Reason in Kant's sense of the term. However, in accordance with Kant's own doctrine of reason, it is possible to invert the relation once more, and ascribe to the incomplete character of the world that which motivates and justifies the transcendental movement of the search for ends—the movement toward the unconditioned.

There is one tenet of his first *Critique* that Kant has maintained throughout his work, namely, the explicit rejection of any hierarchical sense of being, as established by a relative subordination of the elements of knowledge, intuition and thought. For this he wants to substitute their difference of origin or absolute heterogeneity (A844/B872). The basic elements of the natural world itself (the forces of attraction and repulsion) are not exempt from this heterogeneity at the origin. Furthermore, what we can say about knowledge and being, we can say about the history of pure reason, which cannot undergo any dissolution; for otherwise it would fall back on the religious concepts that prevailed when philosophy was in its infancy (A852–55/B880–83). Whether in the object of knowledge (sensible or intellectual), in its origin (empirical or noological), or in its method (naturalistic or scientific), we always encounter some internal tension bordering on outright conflict, which determines the true animating force of philosophy as a battlefield.

Now, the most general mode of opposition, which at one stroke connects essence and history, is that of truth versus illusion. Kant insists on the absolute equality between the two elements of knowledge, since all knowledge results from their union (A51/B75). Yet, a dissymmetry appears in this harmonious union in the case of separation, since "thoughts without content are empty" whereas "intuitions without concepts are blind." An empty thought is thus non-sensical, but strangely enough an intuition deprived of the support of the concept preserves a sense, albeit a very peculiar one: for it is not only blind, but also has the power of blinding, as in the case of the transcendental illusion. For Kant such an illusion is altogether the contrary of a thought that comes to nothing. It results from a subreption by which what carries sense in the subjective part of knowledge "presents itself to reason, nay, forces itself upon reason, as being objective" (A792/B820). Kant diagnoses such a mistaken objectification of the conditions for the possibility of any knowledge as an unobserved influence of sensibility over the understanding (A293ff./B349ff.). This influence is certainly not arbitrary, since Kant conceives of it as inevitable inasmuch as it gives rise to dialectical reasonings—a type of experience which reflects the desire for completeness found in reason.

Dialectical reasonings are thus something like an *amplification of the world* beyond the realm of what empirical experience could possibly invite us to see, just as all a priori knowledge that falls within the range of transcendental truth rests on synthetic, that is, "ampliative principles" (A9/B13). The transition from the empirical to the trans-empirical (which Kant calls transcendent) employment of our concepts is not effected from without. Rather, it is secured by the categories themselves, which *seem* to bear on noumena because they "seem to have more meaning and content than can be exhausted by their merely empirical use."[34] The categories withdraw from the course of appearances in order to serve as their ground; and the ideas of reason then twist a withdrawal into a transgression. Kant's project thus links up with Pascal's: to render reason reasonable, to bring reason to bear on reason. Reason's natural state of war has to be transformed into one of perpetual peace. This transformation is not merely a definitive elimination of conflict. It is akin to a reduction in Husserl's sense, which puts validities out of play, precisely because reason always retains something of its original condition—it is in itself haughty and quarrelsome. Husserl's seeming lack of firmness, when he has to decide between the self and the world, makes light of the imminent war that threatens us at all times. The appearances viewed through reason's magnifying and distorting glass are thus the first example of a sort of transcendental reduction of the ego, a reduction which has the effect of liberating the potentialities of the world that remained hidden in consequence of the Copernican revolution.

CHAPTER 3

An Experiment with Concepts

What are the Objects of Reason?

If we look at Kant's table of categories, we see under the first heading (quantity) that the totality is the unity of a plurality, that is, it provides the synthesis of these two categories. What happens to totality when the plurality to be united is itself undetermined? Central to Kant's preoccupation with cosmic concepts in connection with laws of nature is that species of transcendental ideas that refer to the "absolute totality in the synthesis of appearances" (A407/B434). Kant calls this species of ideas "cosmological ideas." Precisely because they never leave the field of appearances, even though they push (*treiben*) the synthesis so far that it reaches a point beyond any possible experience, cosmological ideas are torn apart by a tension that only an abatement in the thing in itself would resolve. Such abatement is not permitted in the case of the cosmic concepts, which is precisely what bestows upon them the "dignity" proper to any representation of ours that has a necessary relation to an object of experience (A197/B242). It is only if we are ready to live with this tension, and carry the effort up until its ultimate implications, that we ourselves will become worthy (*würdig*) of our own reason. The temptation to resort to the thing in itself as a simple expedient is a threat that never ceases. Human reason, Kant writes, "proceeds impetuously, driven on by an inward need, to questions such as cannot be answered by any empirical employment of reason" (B21). Transcendental philosophy does not aim at abrogating this natural desire for metaphysics, which always expresses itself with impetuosity. There is nothing that could ever prevent us from representing "the world . . . as having originated from an idea" (A815–16/B843–44). But this representation needs to be evaluated in the light of the prospect of deploying itself in harmony with a system of ends. That is, the effort involved in holding the cosmological ideas does come to an end when that which is demanded so impetuously from the outset by human reason is satisfied, namely, a moral employment of our faculties and the absolute necessity of the one primordial being. But as for the effort itself, it gives a direction (*Richtung*) to the new investigation of nature carried out in conformity with a system of ends, the principle of which is the systematic unity contained in "the ideal of supreme ontological perfection."

Viewed from the standpoint of this continuing effort rather than from that of the end result, the criterion of perfection undergoes an important shift in meaning. Kant borrows the term "transcendental" from the scholastic tradition, where it refers to all that can be ascribed to an object independently of any particular determination, such as being, unity, and so on. General logic, which abstracts from the content of knowledge, provides the direction to be followed by the understanding in establishing the transcendental content of our representations (the pure a priori concepts of understanding): this direction is the unity given by the function of thought in any *judgment*. A judgment is a substitute for the lack of a faculty of *intuition* that would be incorporated in the understanding: whereas an intuition relates to an object immediately, the judgment, as a combination of many representations, can do so only mediately, by which it becomes nonetheless a representation *higher* than any immediate and particular representation that may be given through intuition (A68–69/B93–94). Now, transcendental logic assumes this unity found in judgment to be transposable to the mere synthesis of various representations in an intuition (A79/B104–5). And just as the judgment makes up for its lack of intuitive immediacy by transcending all immediacy and particularity, so when an intuition receives unity from the function of thought, it becomes intuition in general, which Kant identifies as "transcendental content." Once it can be admitted that the two movements toward generality (from within judgment and from within intuition) must cooperate in order to produce knowledge of determined objects, the overall transposition can be regarded as having transpired without loss or gain: "there arise exactly the same number of pure concepts of the understanding . . . as . . . there have been found to be logical functions in all possible judgments."

But according to Kant, the Scholastics happened to confuse this direction of thought offered by general logic with the ultimate result of all determination. Indeed, Kant criticizes the Scholastic's incautious transposition of the structure of judgment to that of intuition, their failure to return to the original givenness of intuition in order to determine an object. Their haste to satisfy reason's desire is particularly visible in the way the Scholastics used the transcendental predicates contained in their famous proposition: *Quodlibet ens est unum, verum, bonum*. These predicates were set down with a view to extending the class of a priori concepts of objects, thereby surpassing the categories resulting from the mere interplay of judgment and intuition (B113–5). Kant argues that these predicates turn out to be requirements pertaining again to general logic only; they do not relate to things at all. How could the Scholastics confuse the formal with the material meaning of these predicates? The confusion is that a certain quality of knowledge could pass off for its quantity. The illusion is created and maintained by the fact that such inadvertence seems to repeat in its own way (without loss or gain) the double transposition that characterized the relation between judgment and intuition. Indeed, when the three categories of quantity (namely, unity, plurality, totality) are elevated from the material meaning (which they have in Kant's sense of the transcendental) to a formal meaning, they fail to return anything to the objects of experience. Instead of providing means to think something in each of them, they dilute into general qualities of the knowledge of things in general. Thus in the case of unity, its qualitative meaning reduces it to the kind of unity that could characterize the "theme in a play, a speech, or a story." In the case of plurality, "the greater the number of true consequences that follow from a given concept, the more criteria are there of its objective reality." Finally, in the case of totality, the illusion "consists in this, that the plurality together leads back to the unity of the concept;" this return to unity is supposed to mark its perfection. Taken as a whole, the triad of qualities thus repeats, but merely formally, the process of exchange without net loss that took place between judgment and intuition.

The historical past of transcendental philosophy reveals at one stroke the twofold problem of human reason: (1) Reason has a natural, and thus simply irrepressible, tendency to transgress the limits of possible experience (A642/B670); (2) and yet, at the very moment when it is about to reach its highest aim, reason gets caught in the sphere of empty formality. A missed opportunity for completion or perfection: such is the lesson taught by the history of philosophy, which Kant adopts as an essential feature of human reason. But the reevaluation of the ancient concept of the transcendental that Kant seeks to achieve is itself caught in the same movement from which it tries to escape. Thus, qualitative *plurality* becomes the foundation of the merely regulative, as opposed to constitutive, employment of reason's ideas. Such plurality permits only a drawing near universality by aggregating the true consequences of a principle. This occurs when universality, instead of being given as certain (which would be the case in the constitutive employment of the categories), is recognized only as a problem. Let us turn first to the case of *perfection*; as for the third predicate, *unity*, it will be discussed in chapter 4.

Perfection as mere completeness in knowledge does not complete the transcendental table of categories, that is, it adds nothing to the concepts of objects. How, then, are we to satisfy the "inward need" of human reason to reach completeness with regard to both the form and the matter of judgments? The essence of the problem is already included in the proposed change of viewpoint, by which "we can know a priori of things only what we ourselves put into them" (Bxviii). In this sentence, Kant refers to things as objects of the senses. It is only if the object of the senses conforms to the constitution of our faculty of intuition, and hence to our own concepts, that we can envisage the possibility of knowing a priori. Accordingly, the whole critical method is modelled on what Kant calls an experiment: the elements pertaining to pure reason are to be looked for in *"what admits of confirmation or refutation by experiment."* Given that the experiment is workable within the limits imposed by the objects of the senses, the question arises as to what happens in the case of propositions of pure reason that do extend beyond the possible limits of experience, that is, that look towards those ends that can never be attained by strictly empirical means. Beyond these limits, the efficacy of experimentation should be confounded and with it the whole movement towards a critical standpoint. In fact, Kant sees this question as the decisive test that validates the whole Copernican revolution. In order not simply to reinforce but to demonstrate the universal validity of the new hypothesis, Kant asserts that the *very same* objects ought to be seen from a twin perspective, first as objects of experience and then as objects of pure thought. Were it forbidden to proceed so, "we should be landed in the absurd conclusion that there can be appearance without anything that appears" (Bxxvi–xxvii). Pure thought contains the conditions for anything to appear the way it does, that is, to be visible, but this visible is nothing unless a gaze is there to pick it up in accordance with the sensible conditions of any experience (the faculty of intuition). This gaze is nothing other than reason which, in casting about for its proper object, carves out for itself a "space" of reflection so deep that it should be adequate "for the cognition of things" of pure thought.[1]

How can we make an experiment (in some quasi-physical sense) on objects in the absence of any object to experience? This is the problem Kant worries in the Transcendental Dialectic and which he formulates in the preface to the second edition of the *Critique of Pure Reason* by drawing an analogy between his own work and the revolution that Copernicus had effected in celestial mechanics (Bxvi-xvii). Let us note immediately that the Copernican revolution defines a problem for the faculty of reason, not understanding. We interrogate nature in accordance with the ideas of reason, not the concepts of understanding (A645/B673). What is this problem? The explanation of how the heavenly bodies move leads to some internal contradictions if we suppose that they all revolve around the spectator. Copernicus hit on a happier explanation when he made the spectator turn and the stars stay at rest. To look for movement not in the heavenly bodies but in the spectator, Kant goes on to say (Bxxii note), contradicts the senses, but remains true since the fundamental laws of the motions of the heavenly bodies can be deduced from the transfer. These are Newton's laws which give definitive, apodeictic certainty to what Copernicus had first posed as an hypothesis. The problem of reason can thus be formulated as the problem of the passage from understanding to reason via a contradiction with the senses that has to be resolved in some way.

The distinction between objects of experience and objects of pure thought will turn out to be articulated as the distinction between appearances and things in themselves (noumena). Kant does *not* start by telling us that what was wrong in the pre-Copernican worldview was that our knowledge was taken to conform to objects as they are in themselves. Had he done so, this obvious paradox would follow: if our knowledge were to conform to objects as they are in themselves, we would expect the spectators of the world to turn around these objects so that the world could be mapped out completely, while if the objects conform to our knowledge we would expect the latter to be the only possible center in the universe. On this account, a complete knowledge of all appearances should be ultimately tantamount to the knowledge of things in themselves. (Such an interpretation was adopted by Kant's early disciple, Salomon Maimon.) This is clearly not Kant's intention. The point is really that the distinction betweem appearances and noumena *results* from the analogy with Copernicus. The very notion of a thing in itself is articulated as the result of a demand for inner consistency of the experiment with concepts. That is, the thing in itself is not even the pre-critical (say, Humean) notion of object. It is only in the *second* part of the test, which, as Kant says, yields a counterproof of the validity of the initial supposition (the Transcendental Dialectic), that Kant's own notion of the thing in itself comes into being. As it turns out, it has something to do with the traditional sense of object. How this encounter with the classical problems of metaphysics takes place in the Dialectic is thus *also* a problem entirely peculiar to reason as defined in the Kantian sense. But no matter how "classical" Kant's reference to the thing in itself may look like, the point is to understand the nature of objects from the bare possibility of understanding, as if the objects only contributed to setting this possibility into motion. There must be something *irreducible* in the Copernican worldview for it to lay claim to a new sense of appearance, which indeed makes Kant believe, as he proudly announces at the end of the first critical work, that "the critical path alone is still open" (A856/B884).

Thus, two stages have to be carefully distinguished: the description of the objects *before* the analogy with Copernicus, on the one hand, and the description *after* the analogy has been effected, on the other hand. At the former stage, all judgments concerning the difference between appearances and noumena are, as it were, suspended. Kant only speaks then about the *nature* of objects as understood by us, that is, all thinking that determines objects has to be suspended, so that the nature of our own faculty of intuition can be determined in such a way that it has nothing to fear from the (misleading) immediately visible organization of the world. We are asked to withdraw, not from the world of appearances as intuitively given, but from all thinking concerning this world as already constituted in its own right. This withdrawal will have to be undone later, after the true nature of our faculty has been penetrated and expounded. Now, what enables us to establish the reality of our concepts (to use Kant's own wording in CJ 351) is nothing other than the analogy. In the case of objects within the sphere of our experience, analogies can be merely examples to illustrate our concepts when these concepts are empirical, and these examples are adequate precisely because our concepts are drawn from appearances. But when the boundaries of the given world are transcended, that is, when we venture out onto the uncertain sea of speculation where "no intuition can be given that would be adequate to [reason's ideas]", the use of symbolic metaphors becomes indispensable—there is no other source. Even within ordinary sense experience, our mind's cognitive faculty needs those invisibles without which we would not even recognize a dog as a dog. For the empirical concept 'dog' is equivalent to a rule according to which imagination delineates the figure of a four-footed animal in a general manner (A141/B180) without limitation to what experience actually presents. Within the sphere of experience, I already go beyond it inasmuch as I am able to "intuit" the character of an object that is never present in person to my senses, because it is general. So, in the case of objects which are themselves beyond possible experience, any insight will be gained by analogy, not in the usual meaning of "imperfect similarity of two things, but a perfect similarity of relations between two quite dissimilar things."[2] In the case of the analogy with Copernicus, the two totally dissimilar things that are to be "presented" in some way are the appearances and the noumena; and they have a common relation, inasmuch as they both testify to the mind's dealing with invisibles, which is a common feature of *any* thinking which is not itself intuitive. The main effect of the analogy is thus to supplement thought with a relation to the in-itself, while immediately depriving the object of knowledge from this noumenality. Thought is directed away from "real things in themselves" to, more simply, "real things" (A674/B702). But what is it exactly that we gain by losing the direct access to the in-itself? The answer, which will have to be justified carefully, is that we gain, as Kant argues, the possibility of thinking a relation to the "sum of appearances."

The two brands of thought that determine objects in a world otherwise constituted, and which are to be suspended, Kant calls *dogmatism* and *skepticism*. In failing to realize a proper analogy to help us understand the true relations between appearances and noumena, these brands of thought have also failed to articulate properly the presence of invisibles within knowledge. And that is precisely why pre-critical philosophy has remained caught within a series of irresolvable conflicts that did not even give thought a *direction* by which it could orient itself on the "stormy ocean" (A235/B295) of illusion. For the dogmatic use of reason "lands us in dogmatic [*grundlose*] assertions to which other assertions, equally specious, can always be opposed—that is, in skepticism" (B22–23), as Hume's critique has forcefully revealed. Hume's mistake, however, was to seek how to restrict the understanding without at the same time defining its limits (A767/B795). This amounts to taking each and every thing as an appearance, which would always appear simply because the understanding could not operate at all without seeing; on this account, reason would surrender all a priori power to the spectacle of the world. On the other hand, dogmatism rests satisfied with the pure faculty of reason itself, which has always led to such inextricable contradictions as the question whether the world has had a beginning or whether it is eternal. The critique is an experiment aimed at reuniting reason and the spectacle of the world in accordance with their respective merits. The point is to go so much further than Hume himself in the destructive requirements of a critique, by "subjecting to censorship" (what I called above "suspending"), and thus testing, the understanding *as a whole*. Nature as a whole can be defined in its own right as the sum of the objects of experience (Bxix).

A good deal of discussion has arisen out of Kant's claim to parallel Copernicus. Its appropriatness was first put in question by Bertrand Russell, who said that "Kant would have been more accurate if he had spoken of a 'Ptolemaic counter-revolution,' since he put Man back at the centre from which Copernicus had dethroned him."[3] Later attention was directed to the fact that Kant limits his analogy to the first thought of Copernicus (*mit dem ersten Gedanken des Kopernikus*, as Kant says), thereby seemingly restricting its scope to the motion of the earth about its own axis.[4] On the other hand, attempts to rescue Kant tend to overemphasize his tribute to the scientific revolution as a historical event. Thus it has been argued that Kant did not fully complete in philosophy the revolution which he himself initiated, for he would have been dazzled by the name of Copernicus who, in fact, simply moved the center of the solar system from the earth to the sun without giving any attention to the relativity of all centers: "Kant's Copernican Revolution . . . is coherent when properly understood, and where difficulties arise they are the result of . . . his not being Copernican enough."[5] On the whole, it seems that the label "Copernican revolution" has become a way of describing Kant's originality rather than a way of characterizing his own thought.

There is little philosophical insight to hope from textual evidence about the precise use Kant makes of the analogy, nor is there any point in disputing its appropriatness. The analogy becomes interesting when we look at the kind of problems raised by it, not only as it just stands in Kant's preface, but also in relation to other analogies borrowed from celestial mechanics elsewhere in the first *Critique*. Critical philosophy undertakes to give a "definitive" account of the history of theories of knowledge that failed to measure the presence of invisibles, and so these theories can rightly be accused of beating the air since they "wrestle with their own shadows" (A756/B784). But if the endless fight between dogmatism and skepticism now receives the only acceptable solution by both parties, does not Kant's critique bring the history of metaphysics to a close rather than open a unique road? Interestingly enough, there seems to be only one kind of history that acquires a transcendental status in the *Critique of Pure Reason*: this is the history of astronomy, chiefly represented by the figures of Ptolemy, Copernicus, Kepler, and Newton.[6] In addition, the work of such physicists as Galileo and Torricelli is presented in the preface to the second edition in its transcendental significance, precisely inasmuch as the modern revolution in physics can be understood as a *product* of the analogy with Copernicus. More generally, there seems to be a certain similarity between the question of how to apply the categories of the understanding to objects in general and cosmological considerations that stem from the difficulty of applying Newtonian principles beyond the solar system. This is not to suggest that this parallel in any way exhausts the question of the foundations and limits of transcendentalism, nor is it intended to answer Kant's fundamental question as to why human knowledge cannot be satisfied with a purely theoretical system of objects, that is, why this system has to be modelled on a completed object such as the entire universe. But there are some indications in the *Critique* that metaphorical expression as it relates to astronomical theory and the history of astronomy is possibly the only way of coming to grips with the main thesis of the critical method as an experiment (applicable in any context other than the moral). Kant's metaphor for "critique" is ground, while the metaphor for "dogmatism" is flight (A3/B7; A5/B8–9). The paradigmatic case of dogmatism is represented by Plato's philosophy, since it ascends from the world of senses to a transcendent world without reflecting back on the nature of the invisibles encountered on the way: the ascent takes us from a world of air to an empty space (the world of pure understanding), yet in empty space Plato failed to "observe that with all his efforts he made no advance—meeting no resistance that might, as it were, serve as a support upon which he could take a stand." If the metaphor can be taken at face value, then the critique, in following the same pattern from the world of senses to a world beyond, sets out to retrieve a ground in flight, a resistance to the otherwise inertial (mechanical) motion that the understanding follows when liberated from the sclerosis of the senses. In fact, the somewhat systematic use of astronomical metaphor in connection with the "altered viewpoint" (*Umänderung der Denkart*) may well reveal the extent to which critical philosophy necessarily diverges at some point from a strict enquiry into the basis of this particular science such as Newton's. Thus on reading the arguments exposed in the cosmological antinomies themselves, one would get the impression that the cosmological problem arose, in Kant's view, precisely because the critical standpoint was not adopted in the first place. But it must be borne in mind that prior to both the alleged critical solution and the exposition of the conflicts of reason with itself, Kant made it clear that the Idea of Reason "remains a problem to which there is no solution" (A328/B384). If we start out with the problem itself, then it is possible to see just how and why the cosmical concepts of transcendental philosophy, together with their train of necessary illusions, turn out to be a rather special phase, but no more than a phase, of the experience with invisibles pertaining to every possible knowledge.

We have become weary of both dogmatism and skepticism. We are left with "but one critical question upon whose answer depends our future conduct, namely, *is metaphysics at all possible*?" The radicality of the question requires radical means, because "this question must be answered not by skeptical objections to the asseverations of some actual system of metaphysics (for we do not as yet admit such a thing to exist)." We can only rely on the concept, "as yet only problematic, of a science of this sort."[7] We cannot even use the constitutive concepts of understanding in order to extract the sought-for condition of possibility of metaphysics. Therefore, our starting point will be a problematical concept of the system to be realized. But this problematicity will also be the very character of the idea of reason *after* the discovery of the constitutive concepts of understanding.

The Analogy with Copernicus as a Speculative Starting Point

It should be noted first of all that our knowledge of the stars remains totally unaffected by the altered viewpoint; only the more proximate planets are now known in terms of Newton's science of dynamics. The fact is that the Copernican revolution, as it was implemented by seventeenth-century physics, had two effects: the rotation of the earth on its axis was seen as a purely kinematic effect relative to the stars, while the rotation of the earth around the sun took on dynamic consequences relative to the other planets of the solar system. It was the former effect that Hanson chose to identify as what Kant meant by Copernicus' first thought, but it is only in the latter case, when the spectator moves around a definite centre which is not attached to himself in some way, that an indisputable certitude touches on Newton's laws. So if we take Kant's comparison at face value, what it discloses is that the Copernican worldview transposes a contradiction from within the solar system to outer regions: the contradiction of the senses which accompanies the new viewpoint is a remote one (concerning as it does the stars) and it helps annihilate the more proximate one by which the planets were viewed (in the Ptolemaic system) as following erratic and disorderly trajectories.

None of which goes without a need to reconsider the status of what precisely is being unified in the new scheme. In the wake of the controversial debates over the appropriateness of the Copernicus analogy, Hans Blumenberg's analysis is the fullest and the best. Yet, on the crucial matter of its relation to the entire universe, Blumenberg does scant justice to Kant. Explaining Kant's reasons for limiting the analogy with Copernicus to his first thought, he finds that the most intelligible motive is: "The phenomenal diurnal motion brings the greatest multiplicity into the strictest unity."[8] But this largest manifold (the stars) is the whole problem, and as such it exhausts the relevance of the analogy's implications for transcendental philosophy. Taking the sun as the center of the reference system, the contradiction of the senses (to use Kant's own wording) is removed, with the result that we now have "proof of the invisible force (the Newtonian attraction) which holds the universe together" (Bxxii note). Here Kant consistently speaks of the universe, not just the solar system. The sensible contradiction disappears at the cost of introducing something purely conceptual, that is, an invisible force of gravitation. As we progress from Copernican astronomy to Newtonian physics, we thus fall back on the very structure of ordinary sense perception, which needs invisibles (concepts as rules) in order to operate the way it does. This new relation of sense to concept is developed in the work in which Kant set out to examine the propositions of Newtonian physics from a transcendental standpoint. The argument here is that there is no inherent condition in matter or in space to hinder the force of gravitation from having an immediate and infinite influence throughout the universe.[9] But it is the possibility of this force which remains unintelligible: that is, the invisible part of knowledge has to do with its possibility. The force of gravitation, as well as the force of repulsion, is deemed originary precisely because neither of them can "be derived from any other force, i.e. they cannot be conceived."[10] In striving to understand itself as the possibility of conceiving, the concept loses itself, and so the very condition of possibility of any thinking can never come to the surface and become an object of thought in its own right. Newton's physics makes us *understand* the force of gravitation but we cannot *conceive* it, just as the "concepts of reason enable us to *conceive*, concepts of understanding to *understand*—([as employed in reference to] perceptions)" (A311/B367). "Conceive" here stands in place of *begreifen* or *comprehendere*, that is, to determine the conditioned from the unconditioned—the appearance from its condition of possibility, in contrast to "understand" which dovetails with *verstehen* or *intelligere*, that is, to determine the conditioned from another conditioned. The extension of the force of gravitation to the whole universe is a demand of the understanding in its empirical use (that is, as it relates to the appearances as conditioned). It is the natural extension, to all parts of the universe, of the force discovered in the solar system. But only reason is able to conceive from the unconditioned totality which the universe is. It is thus clear that what is designated by totality cannot be the same thing in both cases. The unifying force of gravitation partakes of both the idea of reason and the concept of understanding, however, and it is just at this point that Kant takes leave of Newtonian physics as a well-established body of knowledge in his time and engages with a problem of his own.

Historically Newton's laws were established at the solar scale as a system itself limited in spatial extension. Post-Newtonian science, both in Kant's day and beyond, was confidently based on the powerful heuristic value of the inverse-square law of the force of gravitation, and it was not until the end of the nineteenth century that the chain of inductions (either in the more minute parts of the solar system or at the large scale beyond it) began to break down. What the comparison with transcendental philosophy implies is that the original limitation to the planets of the solar system must be far more than a mere factual limit. In a way, the new contradiction of the senses *is* neutralized when the limit is explicitly marked out and acknowledged as objective, since the contradiction does not then interfere with the agreement between senses and thought already acquired *within* the domain. In comparison with the dimensions of the whole Milky Way, for instance, the movement of the spectator here on earth around the sun becomes quite insignificant.

Newton himself actually ignored the very existence of such a coherent system as the Milky Way. A revealing confusion exists in his own terminology between the words "solar system" and "world."[11] Indeed, no idea of what a physical cosmology can be is to be found in the whole tradition of physical science that the continued successes of Newton's mechanics brought into being. The science established in Newton's *Principia* is a physics that pretends to a certain degree of universality which is gained by a law governing the behaviour of any physical body in any place and at any time. It is a universality by rights, which has no clearcut counterpart in how the matter of the whole universe is actually distributed. Newton himself was quite resistant, since he turned to God when he faced the problem of the stability of the stellar system. This is the famous statement of the General Scholium which concludes the *Principia*: "lest the system of the fixed stars should, by their gravity, fall on each other, he [God] hath placed those systems at immense distances from one another."[12] Of course, the minimization of gravitational effects does not rule out completely the occurrence of an infinite force at infinity. A typical perception of the contradiction that this alleged solution implies, is offered in one of the first popular expositions of Newton's theory on the continent, that of Voltaire, which draws attention to the fact that Newton's theory "seems to contradict itself with regard to the fixed Stars." Indeed, according to this theory, the stars "attract one another, and yet remain immoveable; we must begin with explaining his [Newton's] Sentiment, and shewing, that it does not imply any Contradiction at all."[13] In fact, Voltaire made much bolder speculations than Newton, since he conjectured that "the reciprocal Gravitation of two fixed stars does not diminish precisely in the inverse or reciprocal Proportion of their Distances," that is, "the law of Gravitation may vary."[14]

Such a problem highlights the philosophical aspect of the Copernican revolution which Kant tried to give a secure rational basis. For whether one relies on a (as yet) empirically unsupported variation of Newton's law or on providential action, the idea of universe remains a problem without solution in terms of what is already known with any degree of certainty. There is an obvious clash between the only possible way of postulating a unity of the world, which is that of a dynamic unity, and the perception of a static equilibrium. Newton had once contemplated the possibility that the relative motions of the stars are so slow that they escape notice, only to find a further contradiction, as he put it in one of his letters to Bentley. Bentley must have inquired about the possibility of viewing the general star system as a true dynamical system, rather similar to the solar system itself. If this were the case, the stellar system might perhaps secure its stability that way. But from the comparison with the solar system, Newton finds that "if by . . . any argument, you have proved the finiteness of the universe, it follows, that all matter would fall down from the outsides, and convene in the middle." This gravitational collapse might be eschewed, however, on account of the fact that "the matter in falling might concrete into many round masses, like the bodies of the planets, and these by attracting one another, might acquire an obliquity of descent . . . much after the manner that the comets revolve about the sun; but a circular motion in concentric orbs about the sun, they could never acquire by gravity alone."[15] Newton seems to dismiss the dynamical hypothesis for the whole universe on the grounds that the starry universe is simply not capable of duplicating the existing configuration of a local system such as the solar system. Stars would be bodies constituted like planets, but would move like comets: that is what Newton finds unacceptable. The exemplarity of the solar system is so great that Newton does not imagine movements of stars around a central body as being different from the movements actually exhibited around the sun. Why should it be forbidden to conceive of stars moving only like comets? The answer is that, as Newton has argued repeatedly in the *Principia* and elsewhere, comets in the solar system do *not* exhibit the hand of God: they move in accordance with the natural law of gravity alone, since they are found in every part of it, being not confined to the plane of the ecliptic. Thus, God would contradict his own action if he were to allow a globally dynamical system of stars; the origination of the universe would no longer tally with its mode of conservation.

It is all too clear that the earth is somehow sent back to rest when it is noticed that the dimensions of the observed universe can be enlarged at will. Thus, if the apodeictic and universal legitimacy of the Copernican reversal is to be demonstrated by way of an experiment, as Kant tries to do, the fundamental problem is: what will its status be when and if some purely empirical knowledge starts to be gained from outside the solar system? Where and what will the center be? Of course, Kant would have had no doubts at this point about the progress of science looked at in its purely empirical aspect. In the *Prolegomena*, he is quite eager to show that both in mathematics and in natural science, human reason cannot be in such a position "that it will at any point find completion in its internal progress."[16] This is due to the fact that, in those species of knowledge, "the cognition of reason is homogeneous." *Within* the solar system, the science of dynamics can be said to correct the senses and make them correspond to the kind of ideal accuracy required by the understanding in its proper use. That is why "the sensible world . . . has no subsistence by itself."[17] But *outside* the system, the senses seem to take exception to this correspondence and regain some sort of independence. This is already quite clear from the way Newtonian science thematizes the Copernican system, since it assumes that the relation between the solar system and the stars is purely kinematic. In Kant's wording: through "the mere form of thought . . . I can cognize nothing determinate."[18] Because it loses itself when it deals only with itself in the empty space beyond our solar system, the concept gives way to the senses, and allows them to speak for themselves again. At this point, reason begins to speak the language of irresolvable heterogeneity.

Kant's Cosmological Principle

Kant's way of seeing this issue is associated in his philosophy with a remarkably structured vision of the concrete universe as a totality. This structure takes the form of what we may call the "cosmological principle," a term borrowed from Kant himself.

Before turning to the series of critical solutions to the four antinomies, Kant refers to a cosmological principle whose actual meaning undergoes a gradual and highly significant development (A508–15/B536–43). In the first instance, it is taken in a purely negative sense: "no maximum of the series of conditions in a sensible world, regarded as a thing in itself, is *given* through the cosmological principle of totality." But a moment later, a heuristic value is attached to it: this maximum "can only be *set as a task* that calls for regress in the series of conditions." When it is finally noticed that this exercise itself remains a problem, because it stops us bringing "the regress to a close by treating anything at which it may arrive as absolutely unconditioned," we come to a twofold inversion which Kant sees as the only possible positive meaning of the cosmological principle. First, instead of regarding the world as a whole, and hence as a thing in itself, the principle prescribes a regress in the series of conditions of given *appearances* for which the world as a whole is nothing at all; the task referred to by Kant is thus quite precisely his answer to Hume, inasmuch as it pushes the purely phenomenal character of the world to its largest possible extension. But there is one "given" that gets lost in the process: the givenness of total experience itself. Secondly, and most importantly, the destruction of any given totality whatsoever transforms the regress *in infinitum* into a regress *in indefinitum*. That this means a true inversion rather than a specific change within the hitherto adopted regress, is testified to by the comparison of the regress with the progress. Were we to speak only of the *progress* in a series, the distinction would not matter. Take a straight line: of this line, it may rightly be said that it can be extended to infinity, but it can also be said with equal justice that the advance is indeterminate. In fact, the progress goes from the condition to the conditioned, and so it simply uses a condition of possibility up till infinity; the condition is not changed in its nature, whether or not one wants or must actually perform the synthesis. By contrast, only one member of the series is given in the cosmological problem: the total universe that has been surveyed empirically at any time, by means of scientific procedures, for example. There is no already given line to guide us when transgressing the outermost limit of this territory. However, were the regress infinite, the blind (indeterminate) continuation of the series would again be used; that is, it would be again a matter of indifference whether or not the series had been run through. I would thus transform my present, conditioned situation into a condition, which means that I would be positing my ability to comprehend the whole world at some stage in my thought, thus offering in the last resort the whole universe as the *only* problem for human knowledge. Or else I would have to posit the whole as somehow empirically given: this is the case examined in the second antinomy, the cosmological idea of the totality of division of a whole given in intuition.

In the second antinomy, it is clear that we will always find further members of the series; a necessity resulting from the terms of the problem itself, which is why it is both legitimate and possible to regress through the series of inner conditions *in infinitum*. Not so in the case of the first antinomy. Even if the whole of nature is indeed given, I will never be able to know that it is so given (A482–83/B510–11). Thus, "since no experience is absolutely limited, the necessity is that we *enquire* for them," and so comes the fundamental implication: "However far we may have advanced in the ascending series we must always enquire for a still higher member of the series, *which may or may not become known through experience*" (A518/B546). A statement of this sort is really of the greatest interest if we bear in mind that Kant never tires of insisting on the necessary relation to experience in the constitution of all forms of natural knowledge. In the transcendental deduction of the categories, Kant asserts that the specific character of the laws of nature cannot be derived from the categories themselves. To obtain any knowledge whatsoever of these laws, we must have recourse to experience (B164–65), at least in the case of the so-called special laws; since nature *in general* is defined as the conformity to law of all appearances in space and time, there are also a priori laws that alone can instruct us in regard to experience in general. In the process of the ever renewed enquiry referred to in the positive aspect of the cosmological principle, it is clear that the gaining of "special" knowledge is no longer necessarily related to its transcendental condition, that is, nature "in general." The variance can be understood by playing on a contrast in the figure of the straight line. The linear continuation, the extension of the known parts of the world to the unknown, is the condition of nature in general and thus also of what enables us to single out the elements of human knowledge which pertain to its a priori constitution. This type of linear extension is in fact a blind, spontaneous process since it is the tacit assumption of all truly universal cognition (for example, in the case of mathematical physics in the seventeenth century) that the sample of the universe which it unifies is a sufficiently representative part of the totality, regardless of how far it has really advanced. The cosmological principle put forward by Kant is quite different. It is not, Kant says, a principle of the understanding, in other words it is not dependent upon the objects of the senses enclosed within the limits of already established knowledge, nor is it a principle of reason which would invite us to constitute the knowledge of what lies beyond these limits by way of analogy with what lies within them. For, in the case of both the assertions based on concepts of the understanding and ideas of reason, the precondition of any proof (which is to serve as its guide) is the possibility of experience; no leap (*saltus*) is permitted by which a concept would lead immediately to another concept which is its condition (A783–84/B811–12). Rather, the cosmological principle "serves as a *rule* postulating what in the regress ought to happen from us, but *not anticipating* what is given in itself *in the object* prior to all regress" (A509/B537, translation modified). Thus, passing from nature in general to nature in its concrete totality, the linearity by which no portion of the world can be distinguished from any other is put into question.

What is a rule that does not enable us to anticipate? Reason is the faculty of principles, which are rules for the unity of categories, whereas the understanding is the faculty of rules for the manifold of intuition: the cosmological principle is thus a unique combination of the two faculties. Even though, as a rule, it remains a condition of intelligibility of the manifold, the principle does not really work as a condition. Rather, it is itself transcended by something given, inasmuch as this given is not there yet. Such is the human condition with regard to the cosmos: our action in it (the use of a rule) is also the index of a withdrawal. To be sure, the remotest object of perception is still the index of a possible journey toward it, even if it will never be accomplished in fact. The impossibility for the rule to perform its function of actualization releases a regressive synthesis toward a given which is always put off.

By the cosmological principle, we already know beforehand that the straight line is in fact continuing, but nothing enables us to predict that it continues as the same straight line. In yet another illuminating analogy, Kant compares this to the body in motion as comprehended in modern physics. By itself, the body which follows the rectilinear path is similar to the act proper to the understanding, since this path is the inertial, undisturbed trajectory. But linearity is deflected when, with another force acting in another direction, the body starts off into curvilinear motion. This other force is nothing other than the influence of sensibility. In accordance with this analogy, Kant dislodges the source of the erroneous judgment "as the diagonal between two forces—forces which determine the judgment in different directions that enclose, as it were, an angle" (A295/B352). The task of distinguishing error from truth can be defined, with reference to the geometrical analogy, as the task of resolving "this composite action into the simple actions of the understanding and of the sensibility." In other words, the combination of reason and understanding gives way to a new look at the faculty of sensibility. Sensibility is here endowed with some kind of power over the understanding's tendencies to inertia. In its normal sense, sensibility rather furnishes the purely material part of knowledge, that is, it is a pure receptivity by which sensible impressions may be represented (A50/B74). The ground of error is simply the action of sensibility on the operations of the understanding, and this reverses its being normally subordinated to it. This activity, of course, is precisely what determines reason to extend its conquests beyond the realm of a possible objective knowledge. It enables the understanding to know the member of the series which falls beyond the domain of already established knowledge, *even in the absence* of a member to be experienced as an object. Similarly in natural science, in the Phenomenology section of his *Metaphysical Foundations of Natural Science*, Kant regards the purely inertial state of a body or system of bodies as impossible: "The rectilinear motion of matter in no relation to a matter outside of itself, i.e. such rectilinear motion thought of as absolute, is impossible."[19] Therefore, if it so happened that no experience were possible beyond a given limit, the persisting activity of the sensibility would still induce the understanding to perform a judgment: this judgment, which has deviated from its true function, is the consequence of "the unobserved influence of sensibility on the understanding" (A294/B350).[20]

Kant has just described the way the invisible part of all knowledge becomes sensibly visible, precisely when the business of knowledge is now to transcend its empirical limits. As we saw earlier on when we examined the section on schematism in the *Critique of Pure Reason*, pure understanding could not image anything at all, so that the transcendental schematism of imagination that enables the categories to represent objects could almost certainly never become open to our gaze. But it remained possible to approach the sense of this activity of pure synthesis by relying on the model provided by mathematics, since mathematical concepts happen to be themselves sensible, so that the schematic relation between two wholly different realities (sensible and intelligible) originated from the fact that one reality is contained in the other in an embryonic form. In the faculty of reason, Kant has now found the transcendental (non-sensible) representation of the relation of experience to pure thought that could not be supplied by the work of pure understanding. At this point, human reason begins to gain its absolute autonomy over the mathematical model. There is something to see in sensibility that the understanding in its empirical employment could not see: not the embryonic form of the higher reality aimed at by the faculty of reason, but the residue of activity in sensibility that the original synthetic unity of apperception could not absorb. As it turns out, however, this residue *has* something to do with the sought-for higher reality. The territory of pure understanding is the land of truth, but it is an island surrounded by this wide and stormy ocean, which Kant equates with the native home of illusion. To be sure, if we travel on that ocean, we can be lucky enough to meet an iceberg as sensible signpost, but its hidden part is in fact swiftly melting. By taking our stand on empirical evidence alone, the invisible part of it resurfaces but decomposes itself. In thus losing its original make-up of illusion, we are misled into believing that this method will enlarge the domain of truth. The cosmological problem exposed in the Transcendental Dialectic is concerned essentially with the critical examination of that part of activity which persists in the objects of the senses over and above the newly discovered activity of the spectator. And it is at this level that the cosmological principle sits happily with an actual image of the concrete universe in its structure and appearance. There is nothing coincidental about Kant devoting so much attention to cosmology that, already quite early in his intellectual journey, he could work out a consistent model of his own. In his *Theory of the Heavens*, a work of 1755, Kant tackled, on largely speculative grounds, the true physical difficulties of the Newtonian theory applied to such an object as the entire universe.

The problem Kant had in mind can be phrased in the following terms. If the law of gravitation holds universally, that is in all parts of the universe, the overall attraction of matter will produce a general collapse around some definite center (if the spatial extension of matter is finite) or around an indefinite multiplicity of centers (if the extension is infinite). On the other hand, were nature to provide by some mechanical means an equilibrium between all its parts, the universe would be a simple aggregate of self-contained worlds where "such an exactly measured determination is required in those distances bent by attraction that even the smallest displacement in the universe would entail the downfall of it."[21] One is left at this point with little satisfactory alternative, since a purely mechanical universe seems to rule out any possible form of compromise between motion and equilibrium. To God's miraculous handiwork (one possible alternative), Kant opposes in his early work the idea of a systematic connection between all the parts of the whole by which "one can to some extent picture the system of fixed stars through analogy with the planetary system, if one enlarges this infinitely."[22] The forces which allow the solar system to exist in dynamic equilibrium are those which prevent the planets going into free fall. Kant extends this model to the system of stars which constitute the Milky Way, and even to the systems of such similar systems (to the supergalaxies). In doing so, he was not only the first to identify the actual existence of visible stars as a coherent, dynamic system (the Milky Way to which the solar system is connected as a more or less central part), he was also the first thinker in the whole history of astronomy to intuit both the existence and the systematic motion of extragalactic systems of stars (the nebulae). Furthermore, Kant believes that the true relation of this giant, purely mechanical and apparently self-sufficient universe to God is not in the spatial but in the temporal organization of the world. God's omnipotence and wisdom are most clearly revealed in His ability to create a seeming chaos out of which systematic connection gradually emerges by the power of nature left to itself. That cosmology (the science of the structure of the whole universe) and cosmogony (the science of its origin and evolution) should not be dissociated from one another is actually the burden of Kant's argument in the *Theory of the Heavens*. Within the infinite variety of elements at the first moment of creation, particles of higher specific density were supposed to act as centers of forces, ordering powers by which primeval chaos was gradually organized into a hierarchy of planet-like systems. As a result, "order" is mere locally smoothed-out "disorder" which continually spreads out from some sort of supercenter. The whole of nature is being divided by an ever-receding frontier of achieved order; beyond that frontier, the rest is an infinite chaos that awaits the influence and attraction of the other, already existing forces.

In this dynamic model of the whole universe, Kant thus already faced the overarching issue raised by the Copernican reversal, namely, the possible implications of the motion of the earth to a system *larger* than the solar system. The dynamic model overcomes the obvious contradiction of the senses which the absolute fixity of the visible stars effects. But Kant's point is that the contradiction arises at a further remove. Wondering at how the rest of the universe appears from the earth, Kant notes that it seems "a fully completed world, . . . an infinite host of world-orders which are systematically connected," but that this is our illusion because we "find ourselves actually only in a vicinity to the center of the entire nature where she has already evolved from the chaos and acquired her appropriate perfection."[23] Were some imagined observer to overstep the boundary of order, he would confront the dispersion of elements and the inhomogeneities which make up the primary stuff of the universe. The validity of a postulated homogeneity that would extend universally (on the basis of the actual appearance of the universe in its visible aspect from the earth) is thus just as limited as the perceived analogies between the shape of the Milky Way and supposedly extragalactic nebulae. The perceptions are essentially partial—in the sense that they tell us nothing by themselves about the unity of the universe—until they have been subsumed under a dynamic concept which unites them. But by virtue of its evolutionary character, the unifying concept *forbids* the blind extension of the perceived order on to the as yet unperceived parts.

To be sure, the metaphor of the melting iceberg in the first *Critique* shows how much the critical philosopher no longer believes in the letter of his early model: the melting is an apt image for the vanishing of any postulated organization beyond the island-universe of truth. But it should be borne in mind that Kant never completely abandoned some of the most speculative views expressed in his *Theory of the Heavens*, however dramatic the implications of the later critical philosophy might have been for every aspect of cosmological theory. He refers explicitly to it in certain texts where the spirit of critical philosophy is all too clear. Thus in the *Critique of Judgment*: "The earth's diameter can serve . . . for estimating the planetary system familiar to us, and that [in turn] for estimating the Milky Way system. And the immense multitude of such Milky Way systems, called nebulous stars, which presumably form another such system among themselves, do not lead us to expect any boundaries here" (CJ 256). The celebrated conclusion of the *Critique of Practical Reason* is also striking: "the starry heavens above me and the moral law within me . . . the former begins at the place I occupy in the external world of sense, and it broadens the connection in which I stand into an unbounded magnitude of worlds beyond worlds and systems of systems and into the limitless times of their periodic motion, their beginning and their continuance."[24] Furthermore, however overwhelming his new philosophical task, Kant did not give up extending his theory in keeping with the progress of physical science. A paper of 1785, which is concerned with the volcanoes of the moon, provides a proof of this persistent interest. Prompted by certain new observations and consequent theoretical speculations on the actual shapes and origins of the mountains of the moon, Kant undertook to build a comprehensive theory of the origin of the earth in terms not dissimilar to his early cosmogonical model. He took for granted the new and bold hypotheses of A. Crawford about heat capacities and tried to account for the origin of both the solar and the terrestrial heat in accordance with the ideas of a primeval chaos. The concluding paragraph of the memoir articulates the deepest philosophical sense Kant attaches to the problem of the universe. He formulates the need "to search among the causes of things (*Weltursachen*) as far as it is possible for us to do so and to follow the causal chain in accordance with known laws as far as it extends."[25] This is nothing other than the regress contemplated in the first *Critique*, here associated with a temporality intrinsic to an evolving nature; of "every epoch of Nature . . . no one of them can be shown by direct observation to be the first." Finally, the theme of many universes without end recurs in Kant's latest work, where "the eccentricity of comets, even of fixed stars"—precisely the suggestion that Newton had found unacceptable—"all this seems to constitute systems up till infinity, by progression; but no beginning can be apprehended. Ideas are here dependent upon the faculty of seeing."[26]

Now, in the *Critique of Pure Reason* itself, several passages suggest the importance of the hierarchical picture of the universe. Reason is declared to take the understanding as an object, just as sensibility is an object for the understanding (A662–66/B690–94); reason aims at providing the concepts of the understanding with a systematic and complete unity. It is, of course, impossible to find in intuition a schema for that unity, yet the acts of the understanding deprived of the schemata of sensibility remain undetermined. That is why at least "an *analogon* of such a schema must necessarily allow of being given." In short this is the idea of that *maximum* which was called cosmological principle after Kant himself. In a way that strikingly recalls its use in the *Theory of the Heavens*, the analogy is both the heuristic means for enabling us to carry experience across the factual boundaries and the very sign of the precariousness of any extension on the basis of assumed homogeneity. Indeed, the application of the categories to the schema of reason does remove the undeterminacy of the acts of the understanding; but these acts are not determined in such a way as to yield knowledge of the object itself. Rather, the determinate employment of the understanding with respect to the object lying beyond any and every given experience indicates how "the understanding can be brought into complete harmony with itself." What is, then, this object beyond every given experience which is taken by reason to be the understanding itself—this object which seems to invite the understanding to an awareness of its capacities for comprehending? Or more precisely, in our context of an analogy from which cosmological considerations stem, *where* is it?

Kant compares the frontier of the totality of given experiences with the surface of a mirror. The object lying outside the frontier is just as real as "objects reflected in a mirror" which "are seen as behind it" (A644/B672). It has the inevitability of the optical illusion, in the sense that according to Kant the transcendental appearance does not cease "even after it has been detected and its invalidity clearly revealed by transcendental criticism (e.g., the illusion in the proposition: the world must have a beginning in time)" (A297/B353–54).[27] By inevitable illusion in the transcendental sense, Kant means one that "can no more be prevented than we can prevent the sea appearing higher at the horizon than at the shore, since we see it through higher light rays; or . . . than the astronomer can prevent the moon from appearing larger at its rising, although he is not deceived by this illusion." Elsewhere Kant has tried to give the beginning of a physical and physiological explanation for this phenomenon. He suggests that, seen in daylight, objects which are brighter than the immediately surrounding objects also seem to be larger.[28] An image is produced by a mirror which reflects incident rays, so that the original object appears where, in reality, it is not. Now the optical illusion arises in the case of the source itself not being seen, in which case the reflected image is taken for an object. Thus, the astronomer is not deceived by the illusion of the moon appearing larger at its rising, for he knows how to relate this purely sensible impression to its actual source. The transcendental and the optical illusions have this in common, that there is a certain aspect of the sensible impression which the understanding is never capable of having a hold on: the viewpoint from which the thing is seen; left to itself when transgressing the frontier of the totality of given experiences, the understanding loses completely any sense of a specified viewpoint on things. To be sure, were the astronomer not here on the earth, he would not see the moon appearing larger in certain circumstances, but conversely he would have no way of knowing how such a phenomenon looks like for a terrestrial observer, unless he had a possibly complete *description* of the actual shape of the earth and knowledge of its atmosphere: this description pertains to geography rather physical science, that is, it is precisely no longer dependent on the a priori possibility of knowing anything at all. If we pass on to the greatest possible extension of the understanding (the cosmological problem), "the illusion involved is indispensably necessary if, besides the objects which lie before our eyes, we are also to see those which lie at a distance behind our back" (A645/B673). Objects *behind* us cannot be distinguished from those *beyond* us, since the mirror interchanges the sources and their reflected images.

Of course, the mirror also produces at the same time images of the objects lying before our eyes, and these images are not to be confused by the understanding with their sources. But in becoming its own master, the understanding *is* confusing the two. Because the interplay is necessary, however, the preoccupying center of the critical philosophy of pure reason is the abiding question: "How does reason proceed to set bounds to the understanding in respect of *both* these fields?,"[29] that is, the field of experience and the field beyond it. The perduring activity of the sensibility is simply the process by which the boundary acts as a mirror. Accordingly, reason does not merely *restrict* the understanding, but must also establish the determinate and necessary *limits* of it, for otherwise it would be impossible to see how the anticipation of experience could also be the determination of its *proper* object, that is, how a priori and true knowledge would be possible at all. (At A767/B795 Kant makes this point in relation to his attack of Hume's skepticism.) Precisely because such limits are set by reason and not the understanding, the means exist to preclude this determination from being confused with an apparently true determination of a reflected image. Why and how is that so? Reason is not merely something like a superior understanding, even though they both deal with pure concepts. Thus Kant says of a pure concept that it can be called a *notion* (A320/B377) inasmuch as it has its origin only in pure understanding, not in a pure image of sensibility. Is the idea of reason, as pure concept, just another kind of notion? Kant's answer is most emphatically negative. Reason is not the origin of its ideas since it does not even produce a concept all. Rather, the nature of the idea lies in freeing the concept of the understanding from its natural encroachment upon experience (A409/B435–36), and so the idea remains as no more than a pure image of the understanding itself. That the concepts of reason are sifted out from any relation to their source is the essential point in order to understand why they drag us into inevitable illusions. Each object located beyond the realm of experience that has already been effectively run through is, as it were, invested with a mark by which it can be distinguished from its image. Where does this come from? We know from the "Anticipations of Perception" that the transcendental genesis of any knowledge associated with visible marks is doomed to leave no trace behind its coming into being. In the case of reason, the inevitable illusion arises because the actual viewpoint from which the sum total of appearances (the image of the understanding) is seen can never be located, which is another way of saying that it is unconcerned with the origin of its own concepts. Therefore, what pushes reason to invite the understanding to continue the regression towards full completion is precisely that, by doing so, it succumbs to the belief that, by displaying the supposed origin of the understanding's claims, it leaves nothing behind—that everything is encompassed. Extending the analogy with the geometrical picture, Kant can say: "Our reason is not like a plane indefinitely far extended . . . but must rather be compared to a sphere. . . . Outside this sphere (the field of experience) there is nothing that can be an object for reason" (A762/B790). And the synthetic a priori propositions of pure reason are those by which only the knowledge of "the curvature of the arc of its surface" is sufficient to anticipate the whole radius. The peculiarity of reason's capacity to anticipate is that the spherical image (which is the image of the understanding, and thus also of the origin of reason's ideas) gives the illusion of a trace that can always be picked up again.

Yet, since the limits of possible experience are also limits of simply factual order (the actual part of the world which has been surveyed at any time), the sphere is never perfectly spherical. Again, Kant takes the example of the laws of planetary movements (A662–63/B690–91), and his conception can be articulated in the following way. At first, our experience leads us to the idea that the orbits of the planets are circular. This is the type of experience which had dictated the principles of astronomy for the Greeks: how is the image of perfect circularity displayed by the stars to be "saved" in the case of planets? Even Copernicus was obsessed with it, to the extent of conceiving the heliocentric system as a theory overcoming the Ptolemaic equant, that is, that part of the Ptolemaic theory which was forced to depart from perfect correspondence between center and periphery of a circle in order to preserve at least the constant angular velocity of circular motion. It is an *imperfect* experience, however, as Kepler first showed by detecting deviations from circular shape. Kant describes this revolution in terms of the emergence of fixed laws as a substitute for fixed shape: *all* movements will more or less approximate to the properties of the circle, without the circle as such being ever displayed in nature. Thus, the planets follow elliptical paths, and the comets exhibit still greater deviations in the form of parabolic courses. Extending our conquests still further, Kant goes on to imagine, with a clear but implicit reference to the theories of his own early cosmogony, the case of those comets which "entirely leave our solar system," following hyperbolic paths. These are the bodies which, "passing from sun to sun, unite the most distant parts of the universe." The reality of these hyperbolic paths is purely speculative, of course, since Kant himself says that they are "additions which experience can never confirm." In his *Principia*, Newton says that, according to some, comets move even on hyperbolas, so that "they come from the regions of the fixed stars, and are seen by us in their passage through the system of our planets."[30] But Newton himself believed that comets are parts of the solar system, however far they wander afield. Kant needs them only for the purpose of comprehending the unity of the universe as a projection of the unity of our knowledge. In his model, the infinite number of intermediate degrees change the circle into some form of divergent orbit: by turns elliptical, parabolical, and finally hyperbolic. The first two orbits correspond to reality, with the third representing the need of reason to seek out the fullest possible unity—the latter is an image of the former two. Within the sphere of experience, the understanding thus works to warp its originally circular shape. It is the systematic bending which confers unity on the different forms of knowledge which are acquired by the understanding. But reason shapes this unity to the furthest possible extent. At a certain point, near the upper limit of given experience, the warping is amplified to such a degree that no return to the original sphere is possible. It then remains for reason to metamorphose and duplicate the original spherical figure from sun to sun.

Only the conjecture of an endless cycle of spheres may satisfy the ideal of perfection which human knowledge may be seeking through rational knowledge of the universe in its totality. This hierarchy of spheres is where the boundary stands for the complete sphere of the categorial principles: it is the concrete figure of nature in general, that is, the counterpart in concrete nature of nature in general. Subordinate to this figure and part of it is the boundary which constitutes the totality of the knowledge we gain from the special laws of nature and which is a single, yet imperfect sphere. No single perfect sphere can exist any more than sensible impressions can be sustained in their original purity when they follow the action of the spectator's understanding away from sphericity. Just as reason acts to warp the boundary from without, so the understanding warps all purely sensible impressions from within. Each reshaping works in tandem with the other. That each reshaping has complementary effect on the boundary allows the understanding to acknowledge its own harmony in the object postulated by the idea of reason: an object which, as Kant says, is a *focus imaginarius* (A644/B672).

The Movement and Rest of the Spectator

Kant's concept of human reason as a sphere on the verge of disintegrating places his argument completely outside of the classical foundation of modern natural science. Compare Francis Bacon's investigation of the conditions without which no judgment in natural science could be rightly formed. Bacon identifies our reason with a false mirror, producing a distortion of the nature of things because reason's own nature constantly mingles with it. Bacon likens the mind to a bent and dirty mirror whose surface has to be cleaned and made plane before it can produce true images of nature; only then would the separation of truth from illusion leave no residue.[31]

More generally, what Kant perceives is the necessary limit of all natural science, which, following what is now called by cosmologists the Copernican principle of equivalence between all viewpoints, transcends the apparent organization, however concrete and palpable, of the appearances as directly given to us. This principle, of course, has very little to do with the actual work of Copernicus himself. It refers to the formal implications of the Galileo-Bruno principle of relativity. Thus Galileo does not doubt that, in natural science, "it is better to put aside the appearance, and to use the power of reason either to confirm its reality or to reveal its fallacy."[32] For him the correctness of an observation is *never* in question; and the mistake of the Aristotelians was, in fact, to dispute this correctness and argue about whether or not, for example, heavy bodies in falling down from on high, go by a straight and vertical line to the surface of the earth. Transcendental critique unearths the implication of this argument for human reason in its broadest sense, not strictly limited to considerations borrowed from physics, but also not simply extended from them. Galileo believes that where reason intervenes, situations other than the one suggested by the sensory impression are considered in its place. When different situations clash with each other (for reasons of logic and/or observation), can the power of reason be taken so literally as offering its services as the absolutely ultimate court of appeal? If so, then reason's search for complete knowledge would amount to toying with possible worlds, merely selecting in each circumstance the one that best fits the sum total of appearances and theoretical statements related to these appearances. By contrast, the transcendental critique shows that this "toying with" still does not prevent illusion, because when other situations which go beyond the one actually given are examined, they bring with them *their own phenomenality*, which has the peculiarity of being no longer fully controllable by the power of reason alone: the invisible part of every knowledge has been phenomenalized into deceiving appearances, much like the hidden part of the iceberg that melts as it surfaces again. Exercising the power of reason in relation to imagined worlds is insufficient to prevent our stumbling over an appearance, just like over a wall that cannot be passed. The laws of pure reason are certainly not indifferent to the originally Greek sense of law: *nomos*, which derives from *nemein*, that is, the law as a limit in the sense of wall or boundary line.[33] And indeed, Copernicus stands right at the heart of this speculative issue. In his work, the problem that becomes most pressing after the displacement of the center of the world is whether the stars and other planets really enclose the earth or any other center by means of solid celestial orbs. As it turns out, Copernicus timidly clears the ground for others (Bruno or Galileo), who go so far as to destroy all celestial walls altogether in order to break through complete relativity of all centers.

The emergence of a Copernican perspective was concomitant with the new mathematization of nature, that is, with an emphasis on the formal element of exact knowledge and its deep-seated affinity with the material properties of the world. Yet, however universal and necessary the formal aspect is (and Kant himself shows at length in the Transcendental Analytic that it represents first and foremost the transcendental possibility of modern scientific knowledge), the ultimate foundation of mathematized nature lies elsewhere. For the transcendental possibility is itself subordinated to the rational search for unity in which the actual appearances of the universe cannot be freely exchanged. This is expressed in one of the most remarkable and most profound moments in the *Critique of Pure Reason*, when Kant contemplates the critical problem (the limits of knowledge) in relation to the factual limits of experience (A759/B787). The passage must be studied step by step, since it contains in one paragraph not only the different stages which we have traversed so far, but also the return to the singularity of the viewpoint in terms of the extension from local astronomy to cosmology.

Kant begins with the following statement:

> If I represent the earth as it appears to my senses, as a flat surface with a circular horizon [*als einen Teller*, dinner plate], I cannot know how far it extends.

This is a purely empirical representation, taking my geographical position on the earth as a fixed center. Nothing enables me to say that, in any definable sense of the term, a periphery exists.

> But experience teaches me that wherever I may go, I always see a space around me in which I could proceed further; and thus I know the limits of my actual knowledge of the earth at any given time, but not the limits of all possible geography.

This is a new form of empirical representation, which is now linked to my ability to think since it is an *experience*: thus in the preface to his *Metaphysical Foundations of Natural Science*, Kant opposes the manifold of empirical representation to experience (*Erfahrung*), implying that the latter is a cognition.[34] In this experience, the center which I am moves. Or, rather, the movement of the center is a condition without which purely empirical representations and human thought would remain disconnected from one another. The intrinsic connection of the two, as opposed to their mere accidental encounter, comes from the act of displacement. Here we have a genuinely philosophical case of Copernican revolution, since my own movement is such that I am now in a position to realize the limits of my own knowledge.

> But if I have got so far as to know that the earth is a sphere and that its surface is spherical, I am able even from a small part of it, for instance, from the magnitude of a degree, to know determinately, in accordance with principles a priori, the diameter, and through it the total superficial area of the earth; and although I am ignorant of the objects which this surface may contain, I yet have knowledge in respect of its circuit, magnitude and limits.

By increasing the range of my movements and the distances traversed, I find that I know the limits of any possible geography as well as the limits of my knowledge. The final outcome of such symmetrical equivalence is the discovery that a priori principles enable me to stay where I am, without moving at all. I have thus achieved an image of nature in general, thanks to my own movements, and the discovery of the special laws of nature is a task that can be undertaken from this established position. The movements required for that latter discovery are no longer part of the transcendental condition of the knowledge of special laws, but simply its empirical condition.

And this precisely dovetails with the outcome of the Copernican turnabout. Once I have disclosed the a priori elements of any possible knowledge of a given object, by making myself (as spectator) revolve, I reach a position where my movement becomes immaterial: this is the point at which the transcendental condition withdraws, since it does not overlap *at all* with the empirical condition. Let us now try to take in one glance the *sum* of all possible objects of our knowledge. It is a key which confirms that the change to a state of rest is truly inevitable after all our possible movements have been exhausted. This sum

> appears to us to be a plane, with an apparent horizon—namely that which in its sweep comprehends it all, and which has been entitled by us the idea of unconditioned totality. To reach this concept empirically is impossible, and all attempts to determine it a priori in accordance with an assured principle have proved vain.

Before the whole universe, the spectator is necessarily at rest, both in an empirical and in a transcendental sense. By virtue of his fundamental finitude, none of his movements can ever reveal to him the limits of his actual knowledge. The spectator represents to himself the whole universe very much in the way that the earth represents itself to his motionless senses, before he perceives the limits of his actual knowledge. In both situations, there is an apparent horizon which, through an effect of perspective, contains an accumulation of indistinguishable sensible objects. It is essentially this excess of sensible experience which is responsible for the uncontrolled action of sensibility over the understanding. Nor can the spectator empirically survey the whole universe, for he can hardly envisage, let alone encompass, a complete geography of it. What differentiates the two situations is this: in empirical terms, the boundary of any possible experience can never coincide with a definite term, whereas from the transcendental point of view the universe cannot be subject to a physical limit. In transcendental terms it is meaningless to speak of the spectator being at rest, since it is only his movement that brings into being that part of our knowledge which is a priori. Yet, rest appears to be the inevitable outcome of the movement which is conducive to the transcendental, so that the critical philosopher must abstain from physical cosmology for otherwise recourse to it would transform a temporary arrest into a final one. Or to put it in the terms of the astronomical analogy: the picture of the whole universe, whether it is obtained from the earth or from any other place, would be *immediately* dynamic, so that no vindication could be found for the particular change of viewpoint needed for the establishment of Newton's laws.

It is just this broad analogical use of the 'motion versus rest' paradigm that embodies Kant's conception of the inescapable illusions that lurk within knowledge. The contrast between motion and rest leads to two kinds of illusion. In the first place, were I to maintain the reality of my movement with respect to the whole universe, thus keeping to the Copernican reversal even without empirical condition which would allow its implementation, I would now transform this very movement into a source of unassuageable illusion, since the process by which the objects come into conformity with the faculty of intuition would be reversed. That is what Kant means when he says that the idea itself has only an apparent horizon. The appearances generated by the apparent horizon of the idea are inevitable in the sense that they are warranted by the cosmological principle, which strigently imposes that the regress (the movement of the spectator) should always continue. By virtue of the unobserved action of sensibility over the understanding, the principle invites to see appearances even where they may not be, and so these appearances conform to the transcendental function of thought, which is a synthetic and spontaneous act. But on the other hand, what Kant further describes in the first cosmological antinomy of pure reason is the other kind of illusion, equally inexorable, which is generated by preserving the state of rest. This state is a result (or precipitate) of the spontaneous function of thought in the idea. Such upholding transforms the whole universe (though not necessarily the appearances within it) into something like a super-appearance (Kant speaks of "hyperphysicality"), which in the dogmatic position is quite simply the thing in itself.

Accordingly, because there is no empirical way of ever discriminating between motion and rest when the reference is the totality of appearances, the shift to dogmatism threatens the critical philosopher on the very threshold where he might consider entirely new objects of experience. This, in turn, does not go without jeopardizing the act of legitimating transcendentally the foundations of already established natural knowledge. Thus, the experimental employment of our faculties of cognition with respect to the *time* determination requires a return to the naive, prescientific experience of the world, since "we are unable to perceive any determination of time save through change in outer relations (motion) relatively to the permanent in space (for instance, the motion of the sun relatively to the objects on the earth)" (B277–78). Again, it is our own abode in its indefeasible spatial permanence that conditions the possibility of the perception of any displacement, hence of any time determination also. Since before the whole universe any displacement is reduced to something insignificant, all that remains is the one permanent abode itself as always and already given, which enjoins us to suspend all knowledge by which viewpoints are freely interchanged. To be sure, the main business of human reason is to respond to a call of the pure employment of ideas, in such a way as to never "be satisfied save through the completion of its circle [*ihres Kreises*] in [the apprehension of] a self-subsistent systematic whole" (A797/B825). But in order to fulfill this ambition, reason can make use of natural science only as that minimal curved surface which is necessary in order to determine the entire perimeter. It is essentially this articulation of science to transcendental philosophy that helps account for what would be otherwise an obvious paradox: the cosmological principle invites us to always enquire further beyond what is present, but on the other hand, if the critical examination of the boundaries of reason is to be achieved in some way, then this position should "absolve" us "from all further enquiry" (A758/B786). The two demands can be met at the same time because the Copernican reversal is not simply a change from rest to motion for the central position: there is the additional reference to another center at rest which was the sun in the Newtonian paradigm, but which would keep cropping up elsewhere when larger spheres of the world would be known with equal certitude. It is this reference which manages to leave room for nature's own activity, which is represented transcendentally as an uncontrolled action of sensibility. Through this action, our "constraining nature to give answer to questions of reason's own determining" becomes totally meaningful, even though it is also the seat of the unknowable thing in itself. Only a return to the very sources of all possible knowledge, whether sensible (as in the optical illusion) or transcendental (as in the horizon of the idea of reason), protects the activity of the spectator from the danger of being forever nullified by an opposite temptation to the dogmatic quietude of rest.

From the critical standpoint, the essential feature of the Copernican system is that it does not *totally* abolish the centrality of the observer; equally, the Ptolemaic system does not represent a one-sidedly dogmatic viewpoint for, as long as we look from inside this system, it is *immaterial* whether the objects conform to our mode of representation or whether this mode conforms to the objects—there are no experimental means that could enable us to decide either way. Through the Copernican revolution, objects conform to thought inasmuch as they are *observed*. What happens when the field of observation is, as it were, bypassed? We do not simply return to dogmatism, because if we did the initial hypothesis to test demarcation between the two types of conformity (objects to cognitive faculties versus faculties to objects) could not be tested. In order to try out the hypothesis, we must connect the two kinds of conformity with two types of objects that can be distinguished only by thought. If our conclusion is that the unobserved objects, that we call things in themselves, are resistant to our way of thinking of them, we will at least be able to hold onto the proof that the demarcation does not apply to things in themselves, but only to appearances. It follows that the test, which is carried out in the cosmological antinomies, does not provide grounds for a clear-cut refutation of the dogmatic standpoint, a nullifying of its very foundation from which the critical solution would rescue us once and for all, because the dogmatic position is already meaningless when taken in isolation, independent of its critical complement.

Furthermore, the ultimate foundation of the difference between nature and freedom might well lie in the foregoing "transcendental reduction" of natural science. For the ability of theoretical reason to anticipate its perimeter on the basis of a small element of the surface is rooted in its relative independence (or even indifference) with respect to the time of life. To be sure, the mere observation of the sky achieved by humanity, from the earliest records to modern times, is not a period long enough to determine the actual position and trajectory of the solar system with respect to the whole universe. But Kant argues, again with a clear reference to his pre-critical cosmological work, that none the less "on the fundamental premise of the systematic structure of the cosmos, and from the little that has been observed, we can confidently infer the reality of . . . a revolution."[35] Now, in the case of the history of humankind, can we speak of a cycle through which the human species would accomplish the hidden plan of nature, so that all the original dispositions contained within humanity would finally be produced in a perfect constitution? Kant's answer is that the exceedingly long time needed to complete such a cycle renders experience incapable of disclosing anything in advance with regard to whether such a process is real. Here, experience only suggests a "feeling" or "hope" that, after many revolutions and much desperation, the universal cosmopolitan situation may be established at last.[36]

CHAPTER 4

Reversing the Order of Time

The Second Birth of Knowledge

Of the scholastic triad of transcendental predicates—unity, truth, perfection—the latter two are doomed to absent themselves when they are conceived as having a relation to an object. This failure gives a measure of the critical enterprise itself, namely, the setting of limits to human reason. Truth and perfection are reinterpreted as merely formal (empty) expressions of plurality and totality, which are categories of quantity. This is articulated in such a way that the complete series of conditions that would enable us to coincide with nature as the maximum of appearances is itself made intelligible by a regulative principle, according to which this maximum always remains a problem precisely inasmuch as human reason constantly seeks to appropriate it. Now, the movement from plurality to totality is conditioned by the fact that it leads back to the unity of the concept with which we started. What is to become, then, of unity in the scholastic sense? What form will it take after being recast in the transcendental mold? Kant says that unity, when properly understood in terms of transcendental critique, becomes a theme, like the theme that one finds in a story. Let us address this issue of unity by first taking a look at the overall unity of the *Critique of Pure Reason* itself, regarded as a discourse with its own theme.

The "Anticipations of Perception" contain and announce the whole plan. This plan concerns the dimension of time. By validating the existence of magnitudes which correspond to the (necessarily hidden) genesis of a manifold of intuition in consciousness, the "Anticipations" provoke a disturbance in the otherwise naturally fixed (because immediately given) order of time. Indeed, the interval that separates the beginning of the appearance itself from the beginning of its actual perception can never be fixed—Kant speaks of the intensive magnitude as flowing magnitude. This process turns out to be the miniature version of a more radical disturbance of time, one that takes place at the global level of any possible experience. We will ultimately be asked to think of a complete reversal of the order of time.

We must first remind ourselves of the essentials of Hume's problem that led Kant to the discovery of transcendental philosophy. To the question whether we have a priori knowledge of causal relations, Hume answers most emphatically in the negative. Observing the effect, we have no way to deduce the cause; and observing the cause, we have no way to deduce the effect. When one event causes another event, three factors are present—namely, the cause, the effect, and also the causal connection between them. However, in scrutinizing situations borrowed from ordinary experience, Hume fails to find the third term—the causal connection itself. Suppose that one billiard ball lies at rest on a table while another moves rapidly toward it. When they collide, the ball that was at rest begins to move. Three things are noticed about the situation: (1) temporal priority (the cause comes before the effect); (2) spatiotemporal proximity (the cause and effect are close together in space and time); (3) constant conjunction (when the experiment is repeated many times the result is found to be just the same as the first time). Hume is careful to point out that a priori reasoning cannot support the seemingly obvious conclusion that the moving ball will produce motion in the one at rest. We can, without contradiction, imagine many possibilities: for example the ball that is initially at rest might remain fixed while the moving ball returns in the direction from which it came. Moreover, no matter how closely we examine the situation, the thing we cannot see, Hume maintains, is the causal connection itself—there is something like a "secret power" by which the cause brings about the effect. If we observe *just once* two events in spatiotemporal proximity, one of which follows right after the other, we cannot tell whether it is a mere coincidence or a genuine causal connection. Returning to Hume's billiard ball example: on the *first* observation of such a collision we would not know whether the motion of the ball originally at rest occurred by coincidence or as a result of the collision with the moving ball. It is only after repeated observations of such events that we are warranted in concluding that something like a causal relation exists. This, according to Hume, goes so far as proving that the causal connection itself is not an observable feature of the situation. If it were an observable feature we would not need to observe repetitions of the sequence of events, for we would be able to state the causal connection in the first instance. As a result, the basis for our judgments about causal relations is simply that it is a matter of custom or habit. When we expect an event to be followed by another, this is merely a fact about human psychology: we form a habit, we become conditioned to expect such from such. Is there any *reason* for supposing that nature is uniform? Establishing nature's uniformity on the basis of past observations amounts to supposing that it will continue to be uniform, which is simply to suppose that the future will be like the past—a flagrantly circular reasoning. It remains entirely possible, so Hume concludes, that nature will not be uniform in the future—that the future need not be like the past—for we can consistently imagine all sorts of other possibilities.

Any thinking meets unsurmountable limits when it seeks to take possession of, and dwell in, that which is outside itself, but the manner in which pre-critical thinking negotiates the treacherous curve which these limits announce differs from that of critical thinking in regard to the direction of time. Kant articulates two ways whereby thought relates to the world outside its sphere, which are as many ways of relating unnaturally to the given order of time. The first is only apparent, because it is not completely independent from experience (B2–3): by using a general rule which is itself borrowed from experience, we are in a position to anticipate the occurrence of empirical facts in their own natural order. For instance, we do not have to wait for a house to collapse if its foundations are undermined, because we know by experience (repeated association in the past, in Hume's sense) that heavy bodies fall. Kant goes so far as to claim that if the rule, according to which all that happens has a cause, is still grounded empirically, then its universality and necessity would be merely fictitious (A196/B241). The second kind of unnatural relation to the time-order of the external world is true, because its foundation is independent of all experience. This is the properly transcendental disturbance. While the first disturbance brings about an incomplete sense of a priori, the transcendental turn forms the basis of an entirely a priori knowledge. We are now asked to withdraw to absolute independence from experience: past knowledge is no longer to be used with a view to announcing the future course of events. How is it possible that claims to objective universality and necessity may arise from within a withdrawal from all experience?

It is, Kant writes, a "peculiar character" of any transcendental demonstration that it sustains a basic paradox. This peculiarity is that, when a proposition, such as that everything which happens has its cause, is taken as principle, "it makes possible the very experience which is its own ground of proof, and that in this experience it must always itself be presupposed" (A737/B765). What is grounded by a transcendental foundation is, in turn, to serve as ground for that very act of grounding. The movement from the ground to the grounded cannot but repeat the inverse movement from the grounded to the ground. If universality and necessity can be found in a proposition, they must express unity in the separation. But if our concepts were derived by following the merely factual progression from particular perceptions to universal concepts, the separation between experience and its ground would be taken as an always already accomplished fact, and consequently only the "first strivings" of our faculty of knowledge as it comes to grips with experience would be accounted for. Not so in the case of the deduction of truly a priori concepts, which require "to show a certificate of birth quite other than that of descent from experiences" (A86/B118–19). When left to itself, the double movement of experience and its ground becomes a coiling-up. An autarkic sphere arises, that combines experience and ground of experience. This is the self-contained sphere of the transcendental a priori, which gives expression to a second birth over and above the natural birth of knowledge from experience. The second birth cannot take place in any assignable time of everydayness. It can only fill the present with a new sense of what is presentified in experience. At this point, the natural series of phenomena in time is, as it were, suspended. Since natural experience has been burdened with its own ground, the mind caught in the transcendental reflection cannot fail to impose a tension on the manifold of intuition. What is it that this tension brings to light?

The categories of understanding (transcendental rules) are deployed as a result of attending to "the spontaneous generation on the part of our understanding" (A765/B793); this spontaneity has a counterpart in the world of experience itself, namely, the spontaneous synthesis of the manifold of intuition. Understanding and sensibility are then brought together, and are thus able to cooperate, because they are two spontaneities. Thus, the self-containment of experience and its ground discloses something like a snapshot of the series of causally-related events, as if no time were allowed to elapse any more between the effect and the causality of its cause (A203/B248). Effect and cause are virtually simultaneous to one another; without this simultaneity, the manifold of intuition could not be unfolded, and there would not even be any succession in the first place, which occurs only because time is elapsing and not because its order is fixed in some way. No anticipation of an experience that has yet to come takes place at this level, since we can no longer dispense with waiting for the future to display empirically the natural sequence of its events. We can only anticipate that some present experience is indeed what it is, since from the transcendental viewpoint, an event that actually happens results from "an empirical judgment in which we think that the sequence is determined" (A202/B246). Thus, thanks to some a priori rule issued from that mental state which has absorbed both experience and its ground, we should be able to place ourselves in the past of an event, so that we can represent as necessary the order of successive perceptions that gave rise to that event.

The contrast between the two senses of a priori is now absolute. On the one hand, the incomplete (because nonpure) a priori leads to a complete reconstruction of empirical experience, since it allows anticipation of events which are yet to come in their own absolutely fixed order; this anticipation opens up a meaningful sense of the future. On the other hand, such a reconstruction is blocked in the case of the transcendental a priori since, over and above this order (*Reihe*), time is still apprehended in its sequence (*Folge*) by imagination which remains free to take this order either backwards or forwards (A201/B246). In attempting to capture the spontaneity of the world as a projection of the spontaneity of the mind, Kant is forced to acknowledge that the appearances are "pliable" enough to lend themselves to such a twist. For in order to be determining with respect to succession, the advance of time must be taken backwards to a point in the past from which it can then progress again (A194/B239), a twist which can be achieved by imagination only and its free play with the sequence of time. Kant claims to have thus solved the problem of how to ground the objectivity of our temporal representations without forcing any of them "to go out beyond itself" (A197/B242). But imagination could not appropriate the sequence of time and manipulate it, without surrendering to the world a character which manifests the irreducible givenness of experience. In Kant's language: the residual pliability in the phenomenal world ensures that our apprehension of appearances is directed toward *conditioned* appearances. It is not until an ultimate but still thinkable unity of the categories is brought about by the ideas of reason that the appearances can be entirely reconstructed from within the transcendental perspective: they are apprehended as *unconditioned*. Here the freedom of imagination is constrained to follow a regressive synthesis of conditioned appearances so as to reach the remotest past. Only the encounter of such an ultimate point, in which all appearances refocus, can determine the unique direction of time toward the future, so that the "natural" overlap of order and sequence can be retrieved. This point is a moment of suspension of a new kind. At the absolute origin the knowing capacities of the mind should cooperate with the spontaneity of appearances, not only as this spontaneity appears, but also as it is in its own being. Indeed, the contemporaneousness of the knowing mind and the appearing world no longer results from a tension imposed by the mind. Therefore, the prejudices against the thing-in-itself (which were contracted by starting off from the immediate manifold of intuition) can now be abandoned: the point where world and mind originate gives our representations license to go beyond themselves, without being hampered by the given phenomenality of the world. What happens to the transcendental a priori when it is forced to mix with being? The answer to this question takes us to the heart of the cosmological antinomies of pure reason.

The arguments developed in the four antinomies are aimed at showing that the reconstruction of immediate experience *remains incomplete even from an absolute origin*. Each party, in any of the four transcendental conflicts, lays claim to completeness in its own way. Kant presents the ultimate failure to reach an unequivocal sense of completeness by using a specifically cosmological sense of a priori, which turns out be just as fictitious as the incomplete a priori. What is a specifically cosmological fiction in the transcendental perspective? This emerges most clearly in the arguments concerning time in the first antinomy, which deals with the absolute completeness of the composition of a given whole of all appearances. In the refutation of the thesis (according to which the world has an absolute beginning), Kant asks us to consider that before this beginning there must have been an empty time (A427/B455). Now, in such a time there is simply no possibility for a present time to be distinguished from any other moment: postulating an absolute origin (that is, a first unconditioned member in the series of conditioned appearances) only exacerbates the privilege of imagination to run freely in the sequence of time. In the refutation of the antithesis (according to which the world is eternal), Kant argues that there cannot be an endpoint to an infinite synthesis, that is, each element of the synthesis is a present which, however, should never have occurred for the synthesis to be actually infinite; in this case, the series of appearances in time (which is unconditioned) is too full for imagination to suspend it. What emerges from both refutations, then, is that there is always a present with respect to which the complete reconstruction of the being of appearances from their very phenomenality vanishes as a possibility. A transcendental illusion consists precisely in the claim that this present could be transported anywhere.

Recall Hume's devastating conclusion. It rested essentially on his view of the "secret power" which, because it cannot be seen, can always act in such a way as to modify the future beyond all possible expectations. It is precisely because *we,* as sentient beings, are conditioned in some way by the external world that the future need not be like the past. But Kant wants to solve the problem of how to ground the objectivity of our temporal representations without forcing any of them "to go out beyond itself." His notion of imagination is different from Hume's in that, instead of enabling us to speculate freely about possible scenarios of events that are yet to come, it simply suspends the present course of events. The most we can impute to nature is that its events (the appearances) are themselves conditioned by *other* events, no matter how heterogeneous they may be among themselves. Thus, whereas for Hume the simultaneity between the causality of the cause and the effect would be an exceptional case, Kant takes it as the general case which reflects the role of imagination in the intelligibility of the situation. This difference alone seems to capture the essence of Kant's critical stance. For such simultaneity is pressed upon us as soon as transcendental schematism is admitted. Hume's disturbing conclusion about the merely psychological habit of connecting events by means of causality derives from his reference to an observation that happened just once. But it is not the observation itself that has to be repeated in order to conclude in favour of a genuine causal relation, since a repetition is already available in the mind as it internalizes the situation by applying the appropriate schemata. Reference to a "first observation" renders us unable to understand how the motion of the ball originated, precisely because there is no such observation that could ever have been meaningful to us in the first place. All we can do is proceed to reversing the order of time: we either have to reverse the priority of observation over the capacity to understand, or, if we want to maintain this priority, we have to push the first observation back to the remotest past , that is, to an absolute origin to which it ultimately owes its phenomenality. Kant's fundamental point consists in claiming that, from the transcendental perspective, the reversal is always double, because the priority of the human capacity to understand, which makes appearances conditioned, leads inevitably to the positing of an absolutely unconditioned, over and above all conditioned appearances. Baldly stated: the future is not and cannot be a transcendental problem.

Time from the Transcendental Standpoint

The unity of thought that underlies the transcendental way of thinking in the *Critique of Pure Reason* has been retrieved by following Kant's insights into the order of time and its necessary reversal.

The fictitious sense of a priori gives us an anticipation of the future inasmuch as the latter can be regarded as concealing a fixed order of events. There is no transcendental substitute for this apparent possession of events yet to come, since the transcendental a priori is limited to capturing the collusion between the world and us in and through the spontaneity which belongs to any present. The anticipation of this collusion takes us to the past that determines it categorially, which, in turn, requires the freedom to reverse the natural order of time. But in so doing, we are also, against our own will, as it were, dragged to a point in the past which is so remote that the appearances we started with are confused with things in themselves; the latter are viewed as concealing the being of the former. Kant's critique expresses one particularly pregnant aspect of the tragedy of the human condition: our own reason prevents us from coinciding with the absolute origin of the world from which it is supposed to emanate, yet the phenomenality of the world drives us toward just such an origin. The relation of any present to the origin marks a point of fracture between reason and the phenomenality of the world.

Moreover, as reason comes to grips with its self-imposed practical necessities, this rift at the speculative level drags reason down to another breaking-point in the inverse direction: from the origin to the present. In the moral order of the world, practical reason too reaches an ultimate goal which it locates at the origin: this is the concept of a unique and primordial being, or supreme good. A self-subsisting cause is conditioned by nothing but itself. But in contrast with the case of speculative reason, taking off from the origin in order to derive the conditions of the empirical application of the supreme good to the present of the world is *not possible at all*, not even as an illusion. No matter how far practical reason may guide us in our attempts to find where obligation lies (that is, up till God) "we shall not look upon actions as obligatory because they are the commands of God, but shall regard them as divine commands because we have an inward obligation to them" (A818–19/B846–47).

However, there *is* a sense in which it is legitimate to speak of the anticipation of events which were *initially* thought to be beyond the limits of possible experience. The essence of mathematical physics is consistent with the "production" of such objects as, say, planets. Thus, in the eighteenth century, the planet Neptune was discovered as the result of a calculation, not a direct observation. The calculation was based on the method of perturbation: it was assumed that Jupiter and Saturn obey Newton's inverse square law, yet Saturn seems to be slightly disturbed from its ideally Newtonian orbit. Neptune was literally brought into existence as an explanatory factor for the observed deviation in Saturn's motion. But a priori, the deviation could have been explained by means of many other plausible hypotheses. For instance, some special property could be ascribed to the body of Saturn; a medium of uneven density pervading all spaces could even account for a whole series of anomalies in the solar system. Of these various hypotheses concerning the existence of objects beyond the present field of experience, some will be verified, others will not. Does the transcendental sense of a priori have anything to do with any such postulation of existence? The danger implied in the "imitation" of mathematical physics by transcendental philosophy is never so great as at this point.

Kant distinguishes his mathematical principles from a series of so-called dynamical principles of pure understanding. Whereas, in the former, the synthesis of the manifold is a synthesis of the homogeneous, the latter combine that which is heterogeneous (B201-2). As it turns out, the quasi-simultaneity of such heterogeneous elements of experience as the causality of the cause and the effect results from the admission of the dynamical principle of the succession in time in accordance with the law of causality—the second among the three "Analogies of Experience." More generally, the prototype of a dynamical principle remains none other than the cosmological principle which, as mentioned earlier in chapter 3, is a rule that has the peculiarity of *not anticipating* what is present in the object. In Kant's language this is an example of a regulative principle. Because it broaches the heterogeneous, the regulative principle differs from a mathematical principle which organizes the synthesis of the homogeneous, and which is thus constitutive of the manifold.

The dynamical principles of pure understanding give the metaphysical analogy with Copernicus its definitive significance. The mathematical principles of pure understanding, as transcendental counterparts of mathematical propositions themselves, originated from a change in the *direction* joining concept and intuition—from intuition to concept in the case of mathematics, from concept to intuition in the case of transcendental principles. Similarly, the Copernican revolution in astronomy also marks a change of direction which, however, could not be perfectly symmetrical—from the intuitive spectacle of the heavens as determining the condition of the earth, to the condition of the sun as determining the lawful (nonintuitive) organization of the cosmos. Because they support the transition from essence to existence, from the homogeneous to the heterogeneous, the dynamical principles evince another kind of change: a temporal, not a spatial change.

Indeed, an analogy, in Kant's precise sense of the term in the "Analogies of Experience," is based on yet another radical gulf separating mathematics from transcendental knowledge. In mathematics, an analogy not only expresses but *constitutes* the equality of two quantitative relations; if three members of a proportion are given, the fourth is likewise given, that is, can be constructed. By contrast, an analogy in the transcendental sense is qualitative, not quantitative: "from three given members we can obtain a priori knowledge only of the relation to a fourth, not of the fourth member itself" (A180/B222). The fourth member is not and cannot be constructed: we only have a rule to seek it in experience, without guarantee that the search will bear fruit. The analogical relation is not the construction of a term that is still absent. Rather, it underpins a principle of search, which unites in a unique way a presence (as determination) and an absence (as open search). An analogy is like an anticipation of perception in which the manifold of intuition, instead of speeding up to come to meet me, would have slowed down so as to still leave "a mark whereby it can be detected." In this sense, Kant's analogies are principles of waiting. In the "Anticipations of Perceptions," it was already clear that, if I were somehow able to slow down the process of my encounter with the matter of sensation, all the degrees implied in sensation would be intuitable. This slowing down is now possible because I have transferred this power to a special character of the manifold of intuition. This character is the *necessity* which belongs to the manifold "so far as its constituents . . . *belong to one another*" (B201 note), unlike the mathematical synthesis of the homogeneous which is quite independent of any such mutual dependence.

In the preface to the second edition of the *Critique*, Kant says that reason must constrain nature to give answer to reason's own determining. He refers to constraint as experiment. When Galileo rolled balls down an inclined plane, his intention was to force nature to answer his prior question: what is the rate of acceleration of bodies when we let them drop from a height? By increasing the inclination of the plane and changing the weights of the objects, Galileo thought he could derive the law of free fall for all bodies, that is, what happens when an object falls towards the surface of the earth without encountering any obstacle. In his experiments, free fall is first impeded by obstacles, and then, by gradually removing the obstacles (increasing the inclination) we come at last to an understanding of free fall. If we simply waited for unconstrained nature to answer us, then the direct observation of various accelerations would lead us to the false (Aristotelian) answer, since then it could not be concluded that all bodies in free fall accelerate at the same rate. Newton was able to show that Galilean attraction of bodies toward the earth is a special case of a truly universal attraction that prevails among all the bodies of the solar system. At that global scale, is there also an equivalent to constraint in Galileo's sense? Newton's first law states: "Every body continues in its state of rest, or of uniform motion in a right line, unless it is compelled to change that state by forces impressed upon it."[1] The continuation of rest or uniform motion is the inertial, or constraint-free, state. Clearly, the constraint-free case never exists in a world which is always already trapped in the compelling exchange of forces between all its constituents. Yet the inertial state may or may not be constrained, as if natural forces were merely echoing in their own mode of being the free possibilities of constraint imagined by Galileo.

Is there an absolute point of reference which supports this analogy? Newton says that physical time is "absolute, true, and mathematical" because it "flows equably without relation to anything external." But this equal flow cannot be measured, because by definition there is nothing against which it can be measured. Therefore, only sensible, relative time which we can measure by means of the actual relations of motion is "commonly used instead of true time."[2] Unlike inertial motion, absolute time cannot be viewed as either constrained or unconstrained. The ideal is now completely separated from the real. Newton assumes that, in the case of time, common experience can be a decent substitute for an ideal experience that remains impossible to achieve.

Kant's "First Analogy" speaks about time where Newton's first law speaks about motion: "All appearances contain the permanent (substance) as the object itself, and the transitory as its mere determination, that is, as a way in which the object exists" (A182). Time is substituted for motion; permanence is substituted for continuation. The essence of this double substitution is that, against the idealized world of Newton's first law, philosophy and commonsense *agree* to postulate the permanence of the substance. Our apprehension of the manifold of intuition is always successive and therefore changing, and yet we naturally postulate a permanent object over and above this flux because "in bare succession existence is always vanishing and recommencing." We need, Kant says, "something *abiding* and *permanent*" without which the mode of existence of an object could never be determined as either coexistent or in sequence. What is this permanent thing which according to Kant "expresses time in general" and in which alone makes relations of time possible?

Newton's absolute time is not Kant's time as pure form of sensibility (in the sense of the Transcendental Aesthetics). Time as pure form is immediately given as infinite. This infinity, which transcends any imposition of limits by the concept, does not make it absolute. Indeed, the meaning of "time in general" emerges precisely from the *negation* of the givenness of the time of inner sense by the concept; but the negation of something absolute could only lead to annihilation. The bedrock common to all of the analogies is that time "cannot itself be perceived" (A177/B219). This negative dimension comes out even more clearly in Kant's formulation of his First Analogy in the second edition of the *Critique*: "In all change of appearances substance is permanent; its quantum in nature is neither increased nor diminished" (B224). Thus, the analogies are meant to resolve the paradox of a sensibility that contains, by virtue of its "purity," more than could ever be sensed. Inner sense retrieves its human character thanks to the imposition of a necessary order by means of concepts. Objects are determined in their temporal existence via three modes of time in general, which form the basis of the three analogies. Each of them indicates one of the possible forms of being in time, that is, they each give a certain figure (a mark that can always be retrieved) to the "expectation" (waiting) which overlaps with every given in the present. These figures are (1) that which persists, (2) that which is in succession, (3) that which persists with respect to what is successive. In particular, in the "First Analogy," the action of the concept is to liberate what *in the object* is *more* than we can actually perceive. We presuppose that the object is more than what it actually gives of itself; it is linked to, and overtaken by, something abiding, against which it stands out. This something abiding is the "one time" within which all other local times must take place (A188–89/B232); it is very different from Newton's image of local times as substitutes for the one absolute time that can never be measured. In this way, we are confirmed that thinking redoes what has always and already been done in sensing. But whereas in the Aesthetics the postulation of one unique time was an immediate consequence of its infinity, the same conclusion now results from the impossibility of thinking time in such a way that succession could be ascribed to it. That is, it would be absurd to think different times in order to think succession. The quality of permanence can thus be transferred to something that accounts for this quality in every object of experience. Kant calls this permanent "the object itself," or the "substance as phenomenon."

So far the concept of substance has been merely posited in conformity with common sense. The "Second Analogy," which postulates the law of causality, introduces the whole chain of new concepts needed in order to deduce substance from more primordial concepts. Kant writes: "Causality leads to the concept of action, this in turn to the concept of force, and thereby to the concept of substance" (A204/B249). The argument uses the negative interpretation of the condition of time as inner sense. That time cannot be perceived in itself is now stated as a justification for the inability of imagination to do more than just "connnect two states of things in two ways" (B233). That is, establishing the objective order of two (or more) states is not within the purview of imagination. Imagination freely multiplies the ways of connecting these states; imagination makes connections, but it does so blindly, without regard for an objective relation. What makes up for imagination's lack is, of course, the rule-giving activity of the understanding. But in order to grasp why and how this activity emerges from the fact that time cannot be perceived, we must compare the role of imagination at two different places: (1) at the level of the spontaneous generation of concepts, *before* any bringing into play of the principles of pure understanding which combine these concepts in order to arrive at a certain transcendental knowledge; (2) in the Second Analogy itself, in which the simultaneity between cause and effect prevails. In fact, this simultaneity indicates a return to a form of spontaneity in transcendental knowledge, which had been abandoned because of the primordial necessity of distinguishing mathematical from transcendental knowledge.

Time cannot be perceived: as the medium of all receptivity, it cannot itself be "received" in such a way that a receptive faculty could become conscious of itself in it. As the pure form of sensibility, time is undoubtedly spontaneous, but it is a spontaneity that is completely erased, absorbed in a receptivity, and concealed in it. Do we have more than just a negative relation to time? The answer must be found in another faculty, the understanding. These "two stems of human knowledge, namely, sensibility and understanding, . . . perhaps spring from a common, but to us unknown, root" (A15/B29). There is a sense of spontaneity in the concept of understanding which is an opening to the world, not a concealment. It owes its spontaneity to the fact that it is the positing of a unifying order in the succession of appearances. What happens when this unifying order is applied to the temporal receptivity, which is no more than "concealed" spontaneity? The effect of the application is to complete the synthesis by which concepts are spontaneously generated (B151). This is the figurative synthesis, by which we are provided with what is necessary in order, for instance, to think a line: imagination *draws* it (B154). Thus the concept of succession can be said to be generated spontaneously because it results from a *motion* conceived "as an act of the subject." By contrast, the transcendental schema of an *already existing* pure concept could only provide an *incomplete* synthesis. The production of the concept, however, can be accounted for only if subject to the condition that abstraction be made from the manifold of intuition. We "attend solely to the act through which we determine the inner sense according to its form"; it is assumed here that inner sense yields, by itself, no combination of the manifold.

But now, in the Second Analogy, Kant shows just what property of the manifold itself corresponds to the setting-in-motion of the cognitive faculties. This is the appearance which contains a happening (what Kant calls an event), for instance, my perception of the ship which moves downstream (A192/B237). The event is concomitant with a rule of succession which is responsible for introducing necessity in the order of perceptions, unlike the apprehension of a house which, even though successive, is indifferent to the point at which it begins. To be sure, the apprehension of a house is also subject to a rule. But the application of the unifying order to temporal receptivity has the effect of liberating what was spontaneous in time as inner sense, but which had thus far been concealed in its receptivity. This liberation establishes a dynamical relation to the real, so that the world is transformed into a set of events. Inner sense is now able to discern what it is that, in a given appearance, enables us to represent it as an object distinct from other representations. That which, in the appearance, contains the condition of the necessary rule is the object. To summarize: when the unifying force of the understanding (which provides a rule of succession) has been deployed in time, inner sense makes the world appear by discerning, in the appearance itself, the objective condition that sustains its character as possible event.

The final task is to establish the connection between a possible and an actual event. What happens when the possible event is brought back into the actual flow of time? It is not until a parallel connection is found in the way the subject's faculties and the world's events are set in motion that appearances can appear in their true constraints. The world that appears to us in the form of events presents itself under the following constraint: "appearance never goes back from the succeeding to the preceding point of time" (A194/B239). The constraint is provided by time itself, which flows from the before to the after, but in its own progression it is never arbitrary: "The advance . . . from a given time to the determinate time that follows is a necessary advance." We have to combine two opposites: the irreversible course of time (the concealed spontaneity of sensibility), and the rule of succession (an open spontaneity of understanding). The result of the combination is this: "since there certainly is something that follows [i.e. that is apprehended as following], I must refer it necessarily to something else in general which precedes it and upon which it follows in conformity with a rule, that is, of necessity."[3] Concealment gets the better of openness, because the appearance retrieves its objective course at the price of losing its determinateness: "it does indeed stand in relation to *some* [any] preceding point of time," not a determinate point, because the determinate course of time has already carried us past it. Time is a necessity without rule, and when the rule of understanding comes up against it, it loses its determinateness. How could *any* antecedent, or an antecedent *in general,* serve as a condition that determines the event? Does this not dissolve the necessity that Kant is trying to establish for the rule of causality?

Kant calls to the rescue his principle that it is only the *form* of change, the mere succession of states, which falls within the range of the a priori law of causality, not the content, the state of what undergoes change (A207/B252). Nevertheless, the rule remains certainly no more than hypothetical in the following precise sense. As a prefiguration of the distinction between reflective and determinative judgment in the *Critique of Judgment,* Kant opposes the apodeictic and the hypothetical use of reason (A646/B674). In the former, the particular is derived from a universal which is already certain, so that the particular is determined in accordance with necessary rules. But in the latter, which encroaches on reflection, "the particular is certain, but the universality of the rule of which it is a consequence is still a problem." In the case of the law of causality, the remaining problem is that the universality of the rule is such that the antecedent may be anything; the subjection of the appearance to time has not changed the character of the preceding point of time as being *any* point, that is, it has not determined it. Given the dynamical character of the connection between events, the rule strictly forbids me to reverse the order of events, "proceeding back from the event to determine through apprehension that which precedes" (A193–94/B239), in conformity with the appearance itself which cannot, of its own, so proceed.

When reason enters into the regressive synthesis toward ever remoter conditions in the past, thereby trying to join with an unconditioned, it does not hesitate to proceed back from the event. Cosmology is, as it were, reason's last-ditch attempt to determine what the understanding, in its empirical employment, is unable to determine. Kant sees this as a vice: the perversion of reversed reason. Instead of seeking, by pursuing the path of physical investigation (which, however, can never come to an end) the unity in the global connection of things in accordance with universal laws, it is the reversed procedure which is followed. Because it is hypostatized, the reality of a principle of purposive unity is taken as foundation or origin, the consequence of which is that we then "impose ends upon nature, forcibly and dictatorially." Reversed reason, starting from an origin which is imposed by force, can be deemed a perversion precisely because it is not dissimilar to the fundamental problem with the moral order, in which any reliance on an external origin in order to account for inner obligation is imperatively forbidden. Again, concealment triumphs over openness. In the case of the external relations of nature themselves, "if . . . I begin with a supreme purposive being as the ground of all things, the unity of nature is really surrendered" (A693/B721), since it becomes unknowable by means of the universal laws of nature.

What, then, is the legitimacy of rational cosmology founded upon, if reason will ultimately have to give up its claims to reach a constitutive knowledge of the unconditioned? At first, it would seem that enlarging the horizon of experience only makes more intense the problem of causality in relation to events already included within the horizon. Kant writes: "That there may be inhabitants in the moon, although no one has ever perceived them, must certainly be admitted. . . . To call an appearance a real thing prior to our perceiving it, either means that in the advance of experience we must meet with such a perception, or it means nothing at all" (A493/B521). We cannot help but think that sensible objects exist before we have any experience of them, precisely because we have the consciousness of our finitude, that is, we do not produce or create them. Kant speaks of discovery: these things were there before I discovered them, *even though they may remain imaginary* as long as I have not been able to perceive them. The cosmological problem, which deals with such cosmic events as inhabitants of the moon, is an amplification of the problem of causality of ordinary events, inasmuch as it extends the problematical (non-determinative) character of the past condition from *form* to *content*. How is this extension conceivable? Since objects are appearances and not things in themselves, they have no being outside of perception. There is no such thing as a superior consciousness which could compare the object outside of perception with the object given in perception, so that the demonstration that the former might be prior to the latter is never available. Rather, priority is the structure of perception itself, inasmuch as perception is not a simple relation to its object: while I apprehend the object, I also represent the progress of this perception toward a future perception. When I say of an appearance that it existed before my present apprehension of it, I relate to a past perception in which there must already have been the sketch of the connection that would lead to this present perception—which was then in the future. What exactly is this sketch which bridges the gap between the two aspects of causality—form and content?

The Antinomies as the Life of Reason

On several occasions, Kant expressed the view that the cosmological problem played a decisive role in the journey toward critical philosophy. In a draft for a letter to Marcus Herz, dated 11 May 1781, he declares that he "could have begun from what has been exposed under the title of *antinomy* of pure reason, which . . . would have given the reader the desire to then seek the sources of this conflict."[4] The pivotal assertion of the *Prolegomena*, which

Kant published soon after the *Critique of Pure Reason* had appeared in its first edition, is found in the section referring to cosmological ideas: "This product of pure reason in its transcendent use is its most remarkable phenomenon. It serves as a very powerful agent to rouse philosophy from its dogmatic slumber and to stimulate it to the arduous task of undertaking a criticism of reason itself."[5] Kant was asking his reader to perform the same kind of mental revolution he had himself undergone long ago. In a celebrated letter to Christian Garve of 21 September 1798, Kant protested against all attempts to reconstruct his philosophy in terms of investigations on the existence of God, the immortality of the soul, and so on. He claimed pithily enough: "the antinomy of pure reason . . . that is what first aroused me from my dogmatic slumber and drove me to the critique of reason itself, in order to resolve the scandal of ostensible contradiction of reason with itself."[6] In a fragment dated as early as 1769, Kant had spoken of his repeated attempts to prove propositions and their contradictories, "not for the sake of skepticism but because I suspected an illusion of the understanding and tried to find in what it consisted."[7] This could hardly be taken as evidence that some antinomy was discovered in that year, but in any event these statements testify to the dominant role assumed by the antinomies in Kant's most fundamental preoccupations.

Our reason cannot think the whole universe directly and naively in the same way as any other object whose knowledge rests upon its being a limited *part* of the universe. In this respect, the idea that concerns the world as totality is very different from the other two, the idea of soul and the idea of God (A673/B701). There will always be materialists who simply deny the existence of the soul, as well as atheists who simply deny the existence of God. But can we deny the existence of the universe? Surely not, since then we also deny the existence of the sensible world which is a part of it, and the sensible world is indisputably the source of all natural knowledge. In other words, the idea of universe can neither disentangle itself from the sensible nor can it be fully materialized in it. It is primarily this situation which gives rise to the privileged, antinomical form of reasoning: whether it tries to get out of the sensible or whether it seeks to immerse itself in it completely, reason is confronted either way with a certain deceiving appearance. On the other hand, in the case of the idea of soul, the materialist cannot point to an appearance to make his point, so that in the transcendental paralogism "a purely one-sided illusion" (*ein bloss einseitiger Schein*) (A406/B433) is produced. In fact, the proofs of propositions such as "everything which thinks is simple," or the existence of God, must be unique since they "start always from *one* concept only, and assert the synthetic condition of the possibility of the object in accordance with this concept" (A788/B816). The special character of the cosmological antinomies is that the unique concept of universe will yield two proofs. Kant requests on the part of the critical reader "to devote to this antinomy of pure reason his chief attention, because nature itself seems to have established it with a view to stagger reason in its daring pretentions and to force it to self-examination."[8] The matter at hand is no less than to save sound philosophy from death (A407/B434). There is a natural antithetic of human reason as it relates to the world; reason falls into it without effort because it sees itself as part of the world. (No such antithetic is found in the idea of soul, because the soul is not even the same object when reason applies to it in idealistic or materialistic terms.) The dogmatist can always dig in his heels and hold fast to one or the other assertion, like a madman who has lost the sense of a common world experienced with others. The skeptic can no longer escape anxiety, because he gives a fair hearing to every position; he cannot avoid a plunge into despair, an attitude that Kant calls "the *euthanasia* of pure reason." The effort required in order to save philosophy from death is the need to translate the natural antithetic in terms of an actual conflict governed by laws of reason.[9]

Because reason examines itself in order to extract laws from within itself, instead of simply greeting these laws, the cosmological antinomy is the place where the innermost depths of our humanity manifest themselves. In the antinomy, nature speaks to our inquiring minds in the most direct possible way, precisely because, as a complete whole, it is exposed to the danger of being lost in obstinacy or despair. The conflict internal to our reason not only led Kant from dogmatic to critical philosophy, but was responsible for arousing the whole human species from barbarism to civilization. Were it not for a special feature of the human condition that Kant idolises when he considers the beginnings of civilization, namely, a certain disingenuousness in human nature, there would have been no polemic of pure reason in the first place. This is the inclination that we have "to conceal our real sentiments, and to make show of certain assumed sentiments which are regarded as good and creditable" (A748/B776). The essential character of this "outward show of respectability, honesty, and modesty" is that it is literally impenetrable to the gaze of others. At this point, concealment becomes the condition without which openness could not even be thought. Indeed, in so trying to represent themselves better than they actually are, the savages mutually supported each other, became part of a school of self-improvement. Civilization emerged from the interplay of many such fair dissemblances, which are as many conflicts between the internally bad and the externally good aspects of human being. As such, the beginnings of humanity could be only a provisional arrangement. Enlightenment called for overcoming the fundamentally conflictual nature of the beginnings of civilization: the point is now to abolish the old situation (defined by Kant as dogmatism), in which "a righteous cause [was] defended in an unrighteous manner" (A750/B778). In the age of reason, self-control as external appearance can only be spoken of by analogy (A840/B868). The two parties in the antinomy are then like two persons seeking the means for a just and durable peace; in this search, the field of critical philosophy is opened up. The fact, however, that the antithetic, viewed as a *natural* phenomenon of human reason, should be central to the critical project, is itself a paradox. This is because Kant does not address the issue of intersubjectivity in the transcendental context of cognition, except here, where a discussion is *forced upon* two parties in order to alleviate the devastating consequences of a permanently conflictual situation.

In the antinomy, one can thus read the interplay of two dimensions of history: on the one hand, the history of human nature emerging from the state of nature, and, on the other hand, the history of metaphysics emerging from dogmatism. To be sure, human beings have a natural disposition towards metaphysics, but in the state of nature belligerents can secure their assertions and claims only through war (A751–57/B779–85). Given the mischievous tendency to pretend what is not, their claims could then be only false arguments. Does the critical state bring about a peace in the form of a standoff, as if it could content itself with something like a "principle of neutrality" in all controversies? Certainly not, because in such skeptical use of its powers, our enlightened reason would be limited to boasting its own ignorance against the unlimited pretension of its opponents. Once enlightened, reason is bound to enjoy the extravagant privileges of the emancipated slave—something like the state of mind of the *nouveau riche*. As Kant pus it: "Whenever I hear that a writer of real ability has demonstrated away the freedom of the human will, . . . before having opened [the book], I am perfectly certain that he has not justified any one of his specific claims." The humble confession of ignorance, which immediately results from inspecting the antinomy, procures a blissful innocence: it is precisely this that Kant refers to as the awakening from dogmatic dreams. The renewed innocence is thus only the preparatory step for the more pugnacious and mature strategy, without which nothing would remain to raise a conflict within reason—the conflict which will keep us awake and vigilant, for ever, after the advent of enlightment. In the era of critical thinking, the obstinacy of the dogmatic thinker cannot be dispensed with; the dogmatist still contributes to further progress away from savage rudeness, because "he affords me a welcome opportunity of applying and . . . amending the principles of my Critique." The essential contribution of civilization, which is reflected in the critique, is that the disputes between two parties have become "bloodless contests," so that the existence of an antinomy not only ratifies the definitive exit out of the unjust and violent state of nature, but it is also the condition of continuous progress without which humanity would fall back to its earlier barbarous condition.

Thus Reason is the product of history through which humanity unveils its mask of perversity. The whole movement of the *Critique of Pure Reason*, which reaches its climax in the antinomy, is encapsulated in a history of pure reason (A852/B880). At first humans were preoccupied with the hope of a future world. Metaphysics followed as a way of de-theologizing these questions and answers; it brought the lofty preoccupations with an afterworld down to the level of the terrestrial matters of our here-and-now existence. What is the good life, and how it is possible to live it? To Kant's account of the story, we might add that modern science, in the wake of metaphysics, constituted a renewed effort to de-theologize questions which are theological—an effort which can be understood retrospectively on the basis of the renewed tendency of contemporary science to de-metaphysicalize questions which had become purely metaphysical. Clearly enough, the transposition from heaven to earth could not have been effected, had it not been for the fact that modern metaphysics (followed by modern science) was able to absorb, rather than simply eliminate, the first guiding thread of human being's relation to its beyond.

Cosmological theory and speculation lies at the core of this absorption. The Greeks began by articulating the cosmological scheme that lies hidden behind the present order of things. For them, the world must resemble divine perfection as closely as possible. The rotation of the ultimate heavenly sphere (the sphere of the fixed stars) appears to be a true perfection, and therefore, as Plato explains in the *Timaeus* (33b–34a), the world is not imperfect and does not suffer from any deficiency. The search for the true nature of this given perfection sent Plato back to the remotest past, at the time when the universe was crafted by a semi-godlike creature. The universe was produced by a combination of Reason and Necessity: the latter, which is chaos, is represented as a factor in the visible world which confronts the divine intelligence, much like the materials which the human craftsman is given, and which he must use as best he can, though their properties may not be wholly suitable to his purpose. Plato thus found Necessity to be a restricting factor of Reason. No antithetic prevailed between the two, but rather a smooth transition. As he put it: "Reason overruled Necessity by persuading her to guide the greatest part of the things that become towards what is best" (48a). This way of thinking is still present on the threshold of modern science. As promoter of mathematical Platonism at the dawn of modern times, Nicholas Cusanus boldly conjectures that the universe is by itself a mathematical form. This is what he calls in *Of Learned Ignorance* a restricted or concrete form of maximum, and as such it reproduces in the greatest possible way (precisely because it is a maximum) the absolute maximum which is God. For Plato, such imitation was reserved to the demiurge, while here it becomes an inherent property of the world inasmuch as it is mathematical in essence. Mathematical Platonism provided the momentum for the antinomical mode of thinking. Because he advocated the motion of the earth and the stability of the sun, Galileo tried to extricate himself from the authority of the Fathers of the Church by claiming that an antinomical way of thinking is completely absent from their writings. He formulated his defense in the following terms: "Bible texts, their own opinions, and the agreement of all men concurred in one belief, without meeting contradiction from anyone."[10] The stability of the earth admitted by the Fathers *can* be criticized and finally rejected, because "one would have to prove also that they had condemned the contrary opinion." As a result, the antinomical mode of thinking appears to be an exclusive propriety of the specifically modern articulation of the cosmological scheme first offered by the Greeks. Kant's transcendental critique highlights the truly universal dimension of history. Inasmuch as it is a historical product of humans's earliest efforts to situate themselves both in the universe and in their community, the antinomy also includes a trace of the initial impetus that determined traditional thinking to overstep the boundaries imparted to us by sensible nature.

If this broad significance of the antinomy is taken into account down to all its minutest implications, then some positions with respect to Kant's antinomy will turn out to be untenable. The privileged example is the interpretation of the first antinomy. When we are informed by all of the theoretical advances in logic and physics since Kant's day, it would seem that the apagogic type of demonstration used by Kant has lost much of its appositeness. It is well known that, since the work of Cantor, the Kantian concept of totality has become inapplicable when a higher level of generality is considered. It is simply not necessary to enumerate all members of a set in order to maintain a perfectly logical sense of totality, contrary to what Kant argues when he claims that there cannot be an endpoint to an infinite synthesis.[11] Hence, the allegedly rational character of the antinomy assumed by Kant would collapse, even if it could still be claimed that the modern set-theory does not take into account the specifically Kantian relation of concept to intuition as a fundamental premise.[12] In particular, the question as to whether or not Kant's arguments in favour of antinomical structure in scientific cosmology are valid is especially vexing. Nowadays it has become fashionable to deny the antinomies the right of constituting a genuine foundation of cosmological knowledge.[13] The arguments put forward in the antinomy are simply not regarded as plausible in the light of the concrete implications of cosmology, such as those that have been revealed by relativistic cosmology. At the very best, Kant's treatment of cosmological questions is regarded as a rather cumbersome and uselessly complex extension of his general theory of knowledge.

On balance, the temptation is very strong to assign some historically well-defined cosmological doctrine of the past to each assertion in the antinomy. The first conflict of the transcendental ideas deals with the totality of the composition of the appearances into a cosmic totality. On the one hand, the thesis asserts that the world has a beginning in time, and is also limited as regards space, while on the other hand the antithesis asserts that the world is infinite as regards both time and space. In the thesis, Kant would echo the position of Newton, in the antithesis that of Leibniz.[14] Several passages in the celebrated Leibniz-Clarke controversy can be alleged that suggest some similarity between the respective views. For instance, Clarke's argument about space proceeds in the following way: space can only be one of God's attributes, for the very reason that, were this not the case, it would be necessary to state beforehand that the world is of infinite material extension. Now, to assert that God cannot set a boundary to the quantity of matter is, in Clarke's opinion, too far-reaching a proposition to be accepted without supporting evidence, that is, it conveys the ideas of blind fate and pure materialism.[15] Both Leibniz and Clarke lay open conceptions which, to be sure, they do not regard as definitive, but are only consistent with their respective systems of knowledge. Was it incumbent on Kant to give these arguments their definitive structure? Without denying the possible relevance of these and other similarities, they certainly limit our interpretation of Kant without real necessity.

It is not an overstatement to argue that the emergence of Newtonian physics, and its gradual acceptance on the Continent throughout the first half of the eighteenth century, are concomitant with the renunciation of cosmology in the Greek sense of the term—cosmos as order. No substantial theory of the universe as one total system is found, which was explicitly developed upon the foundations of physics set down by Newton and his successors, that is, pioneered and developed systematically before being tested against observation. The Newtonian laws of nature are by themselves independent from any model of the universe. Physical bodies do not even have to belong to one determined cosmic structure, provided that the bodies all obey the same laws. More specifically, the universality of the laws of nature remains unaffected whether the universe is materially finite or infinite; this universality is therefore purely formal. What makes such a conception possible is that these laws are expressed in such a way as to enable us to see nature in true scale, so to speak, from the point of view of an ideally absolute space. This space is purely geometrical and physically inert—it is an arena for physical events, which is itself not affected by the action taking place in it. Consequently, cosmological questioning can be treated only in reference to the formal validity of assertions concerning this arena—not only space but also time, and the nature and form of the laws of mathematical physics.[16] Cosmology in the older sense of the term is then transferred to celestial mechanics, which deals with the astronomical investigation of positions and other measurable magnitudes of the heavenly bodies. Cosmology is reduced to observational astronomy, practically limited to the confines of the solar system.

Alternatively, cosmology in the older sense is also to be found in independent philosophical speculation, renewed by the far-reaching theological implications of Newtonian science. It would be misleading, then, to favour the other extreme and see the Kantian antinomy as one, at best the most significant, of these independent speculations. In this new reduction, which has been taken almost as a matter of course by a majority of Kant's commentators, it appears that Kant is not or cannot be concerned with the epistemological aspect to the question of the foundation of cosmological knowledge. He would tackle this question only indirectly, as a preparatory ground for a very different set of questions: those pertaining to practical reason and the moral order. Thus in section 34 of *Beyond Good and Evil*, Nietzsche speaks of antinomies as the driving power of all philosophical investigations, and then goes on to says that, because of Kant's moralizing turn in his own antinomies, the need arises to go beyond all antinomies, including those of true and false, right and wrong.[17] Of course, it would be impossible to deny that the Transcendental Dialectic is in fact guided, from within its own issues, by the domain of practical reason. In the case of rational theology, for instance, Kant says so quite explicitly, when he expresses the view that "a pure rational faith is . . . the direction sign or compass thanks to which the speculative thinker can orient himself when making rational excursions into the field of super-sensible objects,"[18] as if the absence of the sign would not even permit any excursion in the first place. But it is important to emphasize that this faith, a necessary need of reason, is only capable of "supposing, not demonstrating the existence of a supreme being."[19] The tendency to demonstrate is precisely the main character of reason in its theoretical employment, or rather, reason, in this employment, can only demonstrate. The exclusive authority of the speculative faculty rests upon demonstration, even when the objects are supersensible.

What, then, is the object of demonstration in the Transcendental Dialectic of the first *Critique*? At the very moment when pure reason begins to occupy itself with its practical interest, and so enters into the practical realm in its own right, it is concerned with *two* questions only: the existence of God and the possibility of a future life (A803/B831). Freedom is left out. Indeed, the notion of freedom discussed in the third antinomy is not practical freedom, but only transcendental freedom, which "remains a problem" because, unlike practical freedom, it is completely independent from "all determining causes of the sensible world." Whence transcendental freedom, presented as speculative rather than practical, which seems to be determined by nothing sensible? With our power of knowing a priori, "we are brought to the conclusion that we can never transcend the limits of possible experience, though that is precisely what [metaphysics] is concerned, above all else, to achieve" (Bxix). Consequently, Kant goes on to argue, the plan is "to leave aside the practical ideas, and consider reason only in its speculative employment" (A329/B386). But if speculative reason makes room for its extension to practical reason, it is also forced to leave it empty (Bxxi). Shall we then interpret Kant as saying that only practical reason fills this space, precisely because it alone is adequate to its own concept? This would take no account of the essential character of demonstration related to such an empty space. Beyond the sensible world of nature there is nothing to seize or hold, yet the two parties (represented by thesis and antithesis) involved in trying to take possession of this realm fight because they have brought their shadows along with them: "fight as they may, the shadows which they cleave asunder grow together again forthwith" (A756/B784). A transcendental appearance is a shadow of that sort. It keeps filling a space that remains empty. We avoid it no more than the astronomer can prevent the moon from appearing larger at its rising than at its apogee.

This brings us to those schools of thought that wish to preserve the validity of Kant's arguments. Some commentators have tried to rescue the antinomical scheme, within the contemporary context, by presenting this scheme as the description of alternative interpretations, both equally false, of natural science. The only heuristic value of the antinomy would lie in the way it confronts us with competitive theories within a well-defined program of scientific research. Should one like to save Kant's relevance to present-day science and philosophy, this interpretation would accommodate an avenue for theories that Kant ignored in his own time.[20] Thesis and antithesis in each antinomy would represent the collision of two theories at any stage of the development of science, the collision itself being the ultimate factor of scientific advance. In order for this interpretation to be valid, two readings of the antinomy should be maintained simultaneously. In the proofs themselves, the idea of reason would be regarded as constitutive, so that the ensuing conflicts would be insuperable; but when the idea is brought to the level of a regulative principle, the conflicts would vanish. The integrative interest of reason would dissolve conflicts, a fact that the history of science would allegedly reflect.[21]

Once again, let us not lose sight of the fact that, even before he is concerned with the series of opposite assertions in the conflict of reason with itself, Kant introduces the idea of world as a problem without solution. At the same time, in the practical idea, reason is exercizing causality and brings about that which its concept contains (A328/B384–85). Transcendental freedom is simply the highest point to which the problem without solution can be driven. Thus, if there is anything like a critical solution of the cosmological conflict of reason with itself, it must be appraised in the very terms of the problem. Failing to appreciate this point, all interpretations of the Transcendental Dialectic, whether they aim at discrediting or enhancing the validity of the antinomies, have this much in common: in the antinomies, Kant would temporarily forget about his own critical position, let the dogmatists speak for themselves, and only later would he apply the necessary amount of critique in order to get over the relentless but misleading opposition. A very different perspective, which I shall adopt, is to read the whole argument as the staging of the necessary interplay between dogmatism and criticism. In fact, Kant's argument at that point represents the culmination of an intellectual journey, the consistency of which requires that a certain necessary degree of interaction between dogmatism and criticism has always been with us.

One can think of many objections to the view that Kant does not echo a unilaterally dogmatic conception in the antinomies.[22] If Kant was indeed awakened from his dogmatic slumber by the antinomies, is it not natural enough to believe that Kant formulated these antinomies in a pre-critical form—a form that Kant would simply use again in his Transcendental Dialectic, in order to surmount and abolish it? Does not Kant himself argue that the apagogical type of proof (which is used in the antinomies) does not really strike at the source of truth (even though it can produce certainty), unlike the direct or ostensive proof, which goes down to the ultimate ground of possibility? The apagogical proof suits an object of knowledge which is already subject to a certain constraint, namely, that "it is impossible mistakenly to *substitute* what is subjective in our representations for what is objective" (A791/B819). The constraint is the strongest in the case of mathematics, where such substitution is thought by Kant to be absolutely impossible presumably because in the case of construction of concept in intuition, the matter needs not be substituted for the form—it *is* the form. In this sense, mathematics is dogmatism legitimized by the critique. But the case of dialectical appearance is just the opposite: this is a case where there is nothing to stop the subjective to pass, quite convincingly, for something objective. Should the appearances, as subjective representations, be taken as more than mere images, they would need a *surplus* from the object, for only in this way could they be regarded as having an objective ground. But they are given a *false* surplus of objectivity when, as objects, they are taken as given in themselves (A792–3/B820–21). This is what happens in the case of the whole sensible world, when it is supposed to be given in itself. Whether the totality is finite or whether it is infinite is a fight in the air because both claims are false; truth is completely missed. But does the supposed noumenality result from the use of an apagogical type of proof, or is this supposition the very premise of a thought that leads to its own downfall in the antinomy?

Here we touch on the issue of the problematical status of the thing in itself in the whole *Critique*. In the critical solution, Kant says that the respective proofs are grounded "on the supposition that appearances, and the sensible world which comprehends them all, are things in themselves" (A507/B535). This does not tell us whether the thing in itself is a premise of groundless proofs, or whether the false ground of proofs is revealed as a consequence of the antinomy. In any event, how could the thing in itself be directly thematized in the proofs? Kant concluded the Transcendental Analytic with the observation that a thing in itself is a problematical concept: "it is the representation of a thing of which we can neither say that it is possible nor that it is impossible." It is something like the representation of the irrepresentable, or "the problem unavoidably bound up with the limitation of our sensibility" (A286–87/B343–44). The thing in itself points to what is for us the irrepresentable. Could it be that each assertion in an antinomy toys with one of two facets of that which is irrepresentable? Why, then, should there be just two such facets? The only positive sense of the thing in itself is that it delimits the domain of sensible knowledge in such a way that it fills a space that always remains empty. The representation of a thing in itself serves the sole purpose of opening "a space which we can fill neither through possible experience nor through pure understanding" (A289/B345). Note that the thing in itself is either one of two fillings that come to nothing. As a result, the transcendental appearance which is made visible in the antinomy is the only possible image of the thing in itself, standing somewhere between the purely subjective representation and "the common but fallacious presupposition of the absolute reality of appearances" (A536/B564). But for some reason this representation cannot stay in place; it relapses into either pure experience or pure understanding.

Toward the end of the Analytic, Kant writes, by way of summary of what has been done and what is yet to come: "while the understanding, occupied merely with its empirical employment and not reflecting upon the sources of its own knowledge, may indeed get along quite satisfactorily, there is yet one task to which it is not equal, that, namely, of determining the limits of its employment, and of knowing what it is that may lie within and what it is that lies beyond its own sphere" (A238/B297). The understanding in its empirical employment cannot be determinative with respects to the limits of this employment. Now, what lies or does not lie within the understanding's horizon in its empirical employment is not just objects, but questions related to the very nature of these objects—the problematical character of the thing in itself. If these questions are ignored, then the understanding "can never be assured of its claims or of its possessions, but must be prepared for many a humiliating disillusionment whenever, as must avoidably and constantly happen, it oversteps the limits of its own domain." The overstepping has the peculiarity that it happens constantly, a phenomenon which Kant ascribes to the fact that a determinative principle of limits is missing. The law of nature consists precisely in this: nothing happens without a cause sufficiently determined a priori (A446/B474). No cause is sufficiently determining, in fact some causes are not determined by anything (such is free causality), which leaves room for reflection. How can reflection on the sources of knowledge prepare us for the determination of limits? The answer, according to Kant, cannot be based on a mere "exposition . . . of that which is true," but rather "of that which we desire to know" (A237/B296). It is only by reflecting upon its own desire to know that reason will delimit what lies within and what lies beyond its own sphere. The purposiveness of reflection could not overlap with the being of determination. In the understanding, there is an inner drive that takes it beyond its empirical employment, and this must be mastered in some way by questioning back to the sources of judgment. Indeed, this upsurge, Kant points out elsewhere, is an injunction (or an imperative, *Geheiß*) of the faculty of judging (CJ 188). Let us assume the existence of a principle that would, for once and for all, set the conformity of nature to our cognitive power, in accordance with the "Copernican" mode of thinking which defines the desire to know in transcendental terms. The determinative power of judgment does not provide us with any rule that would enable us to know whether or not limits to this conformity can be detected. The limit is not a particular to be subsumed under a given universal; the power of the determinative rule is limited by its inability to set its own limits. The commensurability of nature with the cognitive power is a principle that has nothing to do with the setting of limits, precisely because it proceeds from the reflective, not the determinative power of judgment. In the determination of limits to which the rational employment of our faculty of knowing is subjected, the universal under which the particular is subsumed cannot be the same as the universal aimed at in reflection.

The reflective sense of universal is found at work in the antinomy itself. The apagogic form of demonstration allows for reflection to try to determine that which determinative judgment cannot determine. Let us begin with the demonstration in the first antinomy, as it relates to space. The argument of the thesis runs as follows. In order to conceive of an infinite universe, we should be able to think of an infinite number of parts of the universe. That is, the successive synthesis of the parts of an infinite universe should be viewed as completed. In order to achieve this, we should be able to count the parts of the universe in an infinite time, that is, an infinite time must be viewed as having elapsed in the enumeration of all coexisting things, which is impossible. Kant concludes: "an infinite aggregate of actual things cannot therefore be viewed as a given whole, nor consequently as *simultaneously* given" (A428/B456, emphasis added). The rule of causality, which locally imposes a simultaneity of all the parts under consideration, is checkmated, from which it follows that the universe must be enclosed within limits. The antithesis demonstrates that the world is spatially infinite. Let us suppose that the world in space is finite and limited, embedded in an empty space which is unlimited. Things will then be related to this outside space. Now, since there is no actually existing correlate to which the world stands in relation, the relation of the world to empty space would be a relation to nothing; such a relation, however, is not itself an object of experience. The rule of causality is again checkmated, since the empty space beyond blocks any search that would allow us to find a new member in the series of objects of experience. Therefore, since a finite world is inconceivable, the world must be infinite in respect of extension. As we have already seen, the argument is similar in respect to time: if the universe has only existed for a finite period of time, then an empty time must have preceded it; but in an empty time nothing could have begun, for in it no one part could be distinguished from any other. In both the thesis and the antithesis, failing to set its own limits, the rule of causality becomes the object of a reflection that seeks the universal in an expanded causality. In the thesis, the reflection on causality comes to nothing, while the antithesis has a blind faith in it.

The notions of space and time used in these arguments are none other than the pure a priori forms of sensibility, not physical space or time. In the observation on the antithesis, Kant even goes to great lengths to show that he is aware of a rather natural objection: a finite world could be conceived of as containing its own boundaries, that is, a limit of the world in space is quite possible without one having to make the impossible assumption of an absolute space extending beyond the real world. In reply to this objection, Kant claims anew that if the sensible world is limited, it must necessarily lie in an infinite void, which admittedly is a nonentity (*Unding*) (A431–33/B459–61). If space as a priori condition of the possibility of appearances is set aside, the entire sensible world vanishes. It would be replaced by the concept of world in general, that is, something purely intellectual that loses any implication for the sensible world. If we continued to assume that Kant is here developing a set of dogmatic arguments, then it would be easy to show that he contradicts himself rather naively, since he could not but borrow elements from his own transcendental doctrine of space and time. It has already been shown conclusively that the substitution of an appropriate physical space and/or physical time to this transcendental doctrine ruins the validity of Kant's demonstration in the antithesis. For example, the notion of non-Euclidean geometry which was developed in the nineteenth century gives the allegedly empty space of the world the status of an object of experience; in this connection, geometry becomes an empirical question, so that the Kantian opposition no longer appears to be a rational structure.[23] The ground for Kant's rejection of both empty space and empty time as physical entities in their own right lies in his assertion that neither could be an object of intuition. But, again, even if the reference to space and time as pure a priori forms were granted, does not Kant stand in contradiction with his own terms, since he writes explicitly in the Transcendental Aesthetics that "we can never represent to ourselves the absence of space, though we can quite well *think* it as empty of objects" (A24/B38–39)? In requesting empty space beyond the limits of a finite world, Kant invites us to adopt at the cosmic level the allegedly disgraced transformation of the sensible world into something purely thinkable (intellectual). Moreover, a similar possibility seems available with regard to time: "We cannot, in respect of appearances in general, remove time itself, though we can quite well *think* time as void of appearances" (A31/B46, emphasis added). More generally, if Kant argues that the apagogic form of proof suits mathematics, why in the antithesis should he refer to the non-mathematical doctrine of space and time which is contained in the Aesthetics?

Similarly for the thesis: it would seem that the role of time has not been fully exhausted. On this point, the following apparently decisive argument has been given:[24] it is perfectly legitimate to *conceive* of a process which takes an infinite time to be performed, and this conception does not, of course, take an infinite time. From a logical point of view, Kant's argument runs as follows: an infinite time that has been running its course up to the present is in itself a contradiction, since an endpoint (the Now) would thereby be posited for something infinite, while according to its concept the infinite can have no such endpoint. Cantor can be regarded as Kant's opponent on this issue. Kant's contradiction arises only because he has already presupposed a particular concept regarding the existence of a whole. According to this conception, such an existence can only be asserted if it is fundamentally possible to present *all* parts of a whole or, as he puts it, to complete the synthesis of these parts. But Cantor has demonstrated the consistent existence of an infinite whole as given whenever there is, for instance, an enumerative procedure employing finite means to assign each part of this whole a particular number belonging to the consecutive cardinal numbers. It is thus seemingly not necessary that all parts be presented. On this view, it is also not contradictory to speak of an infinite whole that has a final, concluding member.

The foregoing criticisms of Kant's arguments are all based on the idea that Kant aimed at constructing antinomies that would be absolutely necessary, not necessary in relation to the specifically Kantian context of transcendental logic. However, there is a way of apprehending the construction in such a way that this distinction vanishes. Space and time are not modified in their character of pure forms of sensibility, because this is the only way, for a synthesis of appearances which is not actually given, to be *allowed* of being given (*dabilis*). Furthermore, because these forms are given as infinite, infinity may be witnessed in Kant's reason. Only the thesis is called dogmatic, since, as Kant says, it rests upon *intellectual* principles (A466/B494). But notice that its demonstration involves the idea that the infinite is unreachable by means of *experience*. That a synthesis could not be completed if it were infinite means that all attempts to think it as completed remain enclosed within the sphere of pure thinking, without any correlate in a real object that I could possibly experience. I am thus obliged to rely on an impossible experience in order to make this thinking part of a possible knowledge. On the other hand, the antithesis is called *empirical*, in the sense that the series of objects of experience fills all parts until the infinite, but the demonstration rests upon my being obliged to *conceive* a relation which turns out to be impossible.

The strength of Kant's arguments then lies in the following mechanism. With respect to the thesis, let us suppose that the impossible experience it rests upon (an infinite enumeration) is annihilated by the possibility of thinking such an enumeration. Instead of destroying the thesis, I then simply fall under the sway of the antithesis: for if I say that the possibility of thinking an infinite time has got rid of the impossible experience of such a time, my own thought would create an experience. This can only be the thought of an experience that is always yet to come: I think an empty time. Now, this empty time eternally fails to make contact with the real universe if no experience other than the thought of something empty can come into it. The principle of sufficient reason (that no point of time in an empty time can be distinguished from any other) is now justified by the attempt to overcome the thesis. On the other hand, the antithesis is the opposite of the thesis only if we begin by supposing that we can dispense with the impossible conception of the relation of something to nothing. If I consider that the experience of the relation of a finite whole to its outside is possible, whether this outside is empty space or nothingness, this does not modify the fact that I then fall under the sway of the thesis. For if I say that by making the relation experienceable I also get over the impossibility of thinking it, I then experience (intuit) something which is an object of thought. But this is exactly the dogmatic way of thinking that Kant does *not* want to use in his own demonstration of the thesis (A430–32/B459–60). The dogmatists first posit the concept of infinity as a given magnitude, and go on to show that such a concept is impossible because no number is the largest—one or several units can always be added to the given magnitude. The specifically transcendental concept of infinity—that the successive synthesis is never completed—is limited to the ever-repeated process of addition, whereas a similar process carried out dogmatically starts from a magnitude which is already unreachable by means of experience. Now, the experience of an object of thought, which would result from the attempt to surmount the antithesis, is precisely tantamount to the dogmatic positing of an experience that begins after the territory of experience has already been left behind.

Therefore, each proof in the antinomy is based on the refutation of the inverse assertion because, in each set of alternatives, we find the very same opposition: namely, the opposition of pure thought to pure experience, or understanding to sensibility each freed from the "natural" relation of one to the other. Because the opposition of pure thought to pure experience takes place *twice*, without our ever being able to say which one actually conforms to its object, we are no longer in a position to state unilaterally that the object conforms to the concept. Taken together, the thesis and the antithesis testify to the fact that transcendental schematism has come to nothing. What appears is an uncertainty about the relation of the object to the knowing subject, an uncertainty which is faithfully reflected in the problematical concept of the thing in itself. In fact, the so-called critical solution to the first antinomy does no more than ratify this uncertainty. The two opposites are not contradictories, that is, they are both false because a third possibility exists, according to which the universe is deprived of any determinate magnitude whatsoever. In the cosmological reasoning about whether the world is finite or infinite, the determination in accordance with magnitude is added only as an accidental (inessential) condition of the world. Only the third possibility reveals the true essence of the cosmological idea, since by virtue of the apagogic type of proof, the arguments of the thesis and antithesis taken together are *already* unable to determine any magnitude. Each proposition only does away with one magnitude, without being able to determine positively the other. As soon as a determined magnitude is ascribed to the world, "we should be regarding the world in itself as determined in its magnitude, and in the opposed judgment . . . attach a determination to the world, regarded as a thing actually existing in itself. This assertion may, however, be likewise false; the world may not be given as a thing in itself, nor as being in its magnitude either infinite or finite" (A504/B532). When I ascribe one of the two possible magnitudes to the universe, I am thinking in such a way that the assertion to be refuted contains a determination that changes the world into a thing in itself. This adjunction of noumenality is precisely what makes me think of it as refutable.

Among all objects of knowledge, the absolute totality thus turns out to be a case of *pure reflection*, that is, its intelligibility is not guided at all by some determination which would always already be lurking. Kant never tires of insisting that, no matter how far experience has been taken, the totality of experience contains limits of its own, limits that no thought can fix. Of the world in its totality, we can only have an Idea. No real object coincides with it: the world as totality of things is, in experience, an object without limit, precisely because the totality of experience is already, in thought, without limit. Within the antinomical conflict of pure thought and pure experience, thought essentially conforms to itself since it fails to have any grip on a sensible intuition that would contain or indicate limits. If a determinate solution to the conflict existed, it would consist in preventing thought from merely conforming to itself. But this is impossible because, even when objects conform to thought within the empirical sphere, thought *also* conforms to itself. Indeed, any possible harmony between received impressions and spontaneous understanding results from the understanding's own capacity to dissect itself into a priori concepts, that is, concepts of the object (A51/B75; A65/B90). To the inward-looking separation of the understanding into its a priori constitutive elements, there corresponds the outward-looking separation of the objects into two classes—appearances versus noumena (Bxxi note).

There is another analogy between the Copernican revolution in its philosophical sense and the Copernican revolution in its astronomical sense which, aptly enough, highlights the nature of Kant's problem (A257/B312). Let us suppose that the sum total of phenomena, inasmuch as it is intuited, can be called the world of senses, whereas the intelligibility of the connection of phenomena in accordance with universal laws defines the world of the understanding. Furthermore, let us suppose that the former world is the world of the ancient tradition (Ptolemy's astronomy), in which the mere observation of the starry heavens is supposed to be directly expressible in theory; whereas the latter world is that of Copernicus and Newton, in which a genuine explanation of observations is proposed. Kant argues that "such a twisting of words is a merely sophistical subterfuge [because] it seeks to avoid a troublesome question by changing its meaning to suit our own convenience." The difficulty is that the world of Copernicus and the pre-Copernican world cannot be characterized in accordance with a simplistic division between a purely intelligible and a purely sensible world. Just as understanding and reason still have an employment in connection with immediately given phenomena, "the question to be answered is whether they have also yet another employment, when the object is not a phenomenon but a noumenon." Now, an object is taken as noumenon precisely "when it is thought as merely intelligible, that is to say, as being given to the understanding alone, and not to the senses." Consequently, it is not just pre-Copernican cosmology which makes the leap into dogmatism. When the earth rotates and the stars are at rest, we are following the injunctions of pure thought; when the earth is at rest and the stars made to rotate, we are following those of pure experience. But in the former case, the agreement of pure thought and pure experience occurs only within our own solar system; if I overstep a certain boundary, I lose the absolute certainty that Newton's law of gravitation continues to apply in the same way. In the latter case, the disagreement of pure thought and pure experience occurs within the solar system—this is the scandal that determined the rise of astronomy among the Greeks, namely, the apparently irregular path of the planets; yet agreement is retrieved at the moment of reaching the unconditioned perfection of the ultimate celestial sphere, that of the fixed stars.

CHAPTER 5

A Logic of Illusion

Reflecting Upon Nothing Determinate: What is a Thing?

The foregoing analogy that Kant draws between the critique and the Copernican revolution is the third and last such analogy in the *Critique of Pure Reason*. The analogy has now given all that it could give, because the premodern distinction between a world of sense and a world of intellect has been completely blurred. The absolute totality of things takes us to the realm of pure reflection: nothing is to be determined. Yet the world of sensible experience is still part of the absolute totality of things; the former cannot collapse just because the latter is undeterminable. How can we have a pure reflection that will yet be determinative in this precise sense? The answer is that pure reflection determines a *schema* of that which cannot be determined.

We have no knowledge of the objective reality of things in themselves. Yet if they are taken as mere thought-entities, nothing prevents us from admitting them "only as having the reality of a schema—the schema of the regulative principle of the systematic unity of all knowledge of nature" (A674/B702). So as not to lag behind the determinate concepts of the understanding, reason undertakes to hypostatize a schema which has no determinate object to realize. This can be achieved only by analogy: "What we then think is a something of which, as it is in itself, we have no concept whatsoever, but which we none the less represent to ourselves as standing to the sum of appearances in a relation analogous to that in which appearances stand to one another." The mutual relations between appearances constitute the world of familiar experience; but the connection between the sum of appearances (the world as totality) and its ultimate foundation is provided by the thought of something, the concept of which can never be determined. Kant argues that an analogy holds between the familiar world and such a world of indeterminate thinking. How do we go by analogy from the relations between appearances to the something in itself which is supposed to contain the ground of their being?

Does natural science, in a sense broader than the Copernican astronomical revolution, throw any light on this ground which could not be gleaned from transcendental knowledge? Reflecting on our radically transformed relation to the world brought about by the developments in contemporary atomic theory, Heidegger believes that the relations between matter and space are no longer as simple as they used to be in classical physics, but nevertheless they are fundamentally the same. What is it that allows us to minimize that impact? The answer is that a feature of space has not really been touched in its foundation: "What keeps a place occupied, takes up space, must itself be extended."[1] However far the division of matter into elementary corpuscles has been achieved, we will never meet anywhere the interior of an extended thing in the pregnant sense of the term: each interior always sends us back to an exterior, without which the relations prevailing among corpuscles of the interior could not be thought. However, because science thinks in accordance with the mode of "how much" and not "what" (the essence), it would seem that this fundamental character of outerness has been passed over. Omnipresent technology, largely as a result of recent atomic physics, has come so close to us that the natural distance between the theoretical thing and the thing of everyday life has burst in pieces. Heidegger wants to restore the distance: "Does one want science even closer to life? I think that it is already so close that it suffocates us. Rather, we need the right distance from life in order to attain a perspective in which we measure what is going on with us human beings."[2] But how could we distantiate ourselves from our own lives, unless something is shared by the familiar world in which we live and the abstract world of science? Heidegger seeks to excavate the truth that mediates between the two worlds. How far can we go in the determination of this mediation?

In order to bring out the common touchstone between the two worlds, we can follow, as Heidegger does, an implicitly Kantian exposition of the philosophical horizon of contemporary natural science. This is Eddington's famous reflection on two tables, the familiar and the scientific. In this reflection, proposed shortly after the formalism of the quantum theory had been set down,[3] Eddington rejuvenates Aristotle's profound observation according to which science deals with universals, not particulars.

The familiar world, he says, is the one "which spontaneously appears around me when I open my eyes." What do I see? I see this table, but by scratching at its surface, I quickly discover that already at that superficial level what is actually meant by such a thing as a familiar table is quite unclear. It has a certain number of easily discernible properties: size, color, shape, and so forth, which will certainly not be described in exactly the same way by everybody; for instance, there could always be some disagreement over the exact color. There are, however, other properties about which agreement may be expected: the number of legs, the kind of material, and so on. If we succeed in distinguishing between these properties which *might* change from those that *could not* change, will we have in any way clarified the question of what we mean by *this* table in particular? Eddington says he does not want to consider how much of a familiar table is subjective (changeable) and how much objective (unchangeable). The point is not that his problem lies elsewhere. His argument is that *any* way of distinguishing the changeable from the unchangeable takes us beyond the table as *this* table. Unchangeable properties belong to a class of tables (having four legs, etc.), while changeable properties do not seem to go much beyond the sphere of private subjectivity. We are led to admit that if, after all, this particular table preserves an identity of its own, then we should be able to define what all familiar tables have in common. But this is also where our powers of investigation seem to give out: we fall back upon the familiar table we started with, because we can hardly say anything more than that the familiar table looks like a familiar table. Consequently, in order to make sense of what the table is as a thing in its own right (or intrinsically), we must postulate something totally mysterious as a constitutive feature: its substance. Substance, in this sense, is totally mysterious because it is not something physically definable in terms of, say, resistance to my leaning upon it. It is substance in the sense of what makes it a thing as such, not a thing with size, color, four legs, and so forth. We arrive at a strange conclusion: substance rescues familiar experience from collapse.

Can we get help by shifting from familiar to scientific experience? Table B is the table as an object of scientific investigation, distinct from its immediate familiarity. We now penetrate the surface a little more deeply. Science tells us about deep-lying atoms and fields of forces. The idea arises that perhaps a complete scientific description of the table will enable us to do away, for once and for all, with the mysterious substance which, apparently, only reflects the limitations of familiar understanding and does not explain anything. What is it that we explain about the table as an *individual* entity when scientific concepts are brought into play? Our additional knowledge has to do with the transformations the table can undergo or be subjected to. For instance, the way it is transformed into smoke under the action of fire. More generally, we know the influences, or connections between things belonging to certain categories of things, including the possible dependency upon measurement and apparatuses of measurement. There is no use in trying to investigate the substance of a thing—what it intrinsically is as an individual entity—by breaking down table A into its more minute constituents or widening its dependency upon measurement. This simply begs the question. The substance of electrons (or quarks, as today's particle physics would have it) would remain as mysterious as the substance of our original and familiar table; the mystery of substance would just be pushed further away from the field of immediate perception. The familiar table owes its original simplicity to its being a strange compound of objective external nature and mental imagery (colors, etc.) which is consciousness-dependent. We thought that for any thing to be granted the status of scientific objectivity in the first place, it would have to be abstracted from what makes it essentially familiar. But precisely by virtue of the strangeness of the originally simple's compound nature, we can never know what "in reality" gets abstracted from what. The boundary between the realm of consciousness and the realm of independent external nature is doomed to remain indeterminate.

We are unable to separate table A from table B in terms of what makes a table just what it is and not something else. This is so at least as far as our *conceptions* are concerned. A difference begins to emerge if we shift our manner of asking about the thing from concept to *intuition*. Let us begin with the spatial features of the table. There is no ambiguity in talking about the familiar table "being there": it stands before our eyes, and therefore it is always seen from a certain perspective. But the scientific table is not merely a perspective of a table; reference to measurement ensures that the changeability due to perspective is eliminated. The scientific table is "really there"—to be sure, not in the sense that as a thing it interacts with consciousness but in the sense that it interacts with another "objective" thing, the apparatus of measurement for instance. However, in the process we lose the familiar understanding of "there." One might be tempted to think: why isn't the scientific table a swarm of atoms located just there, where the familiar one is? This, in fact, is certainly not in dispute. But in the scientific description we are told something about the "reality" of this "there": it is a "there" which is an invariant, not a perspective. This is not so innocuous as it may seem. As Eddington puts it, table B "is really there—wherever 'there' may be." The scientific description can *never* be a description of *this* familiar table because it is impossible to find any trace of the original sense of "there" in the new scientific description. All we can do is surmise that the two descriptions will coincide in all other relevant respects. In other words, the scientific table abstracts from the familiar sense of space just as it removes the illusion of substantiality. In the scientific description, it does not matter whether the table is "here" or "there," since this is totally accidental and cannot belong to the intrinsic features of the thing itself. But Eddington has just said that substance had to be postulated because we cannot know a thing in terms of what makes it what it is. We do not know what "there" may be, and yet we assume that it cannot belong to the intrinsicality of a thing—which is precisely what we do not know either. Just as in the case of the familiar table, we have clarified the question of the intrinsicality of this table only by relying on some external relation: this is the price we pay for abandoning the mysterious substance.

Space definitely plays a strange role. What kind of entity is it, if it is neither a thing nor a nothing? Each table seems to have its own truth, and so what is the truth that mediates between the two tables? Is there something like a third table according to which both table A and table B are true in their own way and represent something like a variation of this superior truth? Isn't the third table holding the key to substantiality? It is precisely because all things are immersed in space, which mediates between the two tables without fixing them in their respective meaning, that we are urged not to prejudge the question of the ultimate identity between world A and world B. The two truths are, for us, so many separate worlds, even though it is obvious that there must be some link between them. In Eddington's words: the world of physics has become a world of shadows, that is, it does have a connection with the familiar world which, however, remains in the dark and cannot be known by any means whatsoever. Space is responsible for this separation; there is no way of recovering either one of the meanings of location (familiar or scientific) by starting from the other. So much so that, in locating a thing in accordance with the scientific sense, we only use symbols (mathematical functions), which have no necessary root in natural spatiality. The upshot of this analysis is that modern physical science is in foreign territory, dealing with things which are ultimately unknowable in terms of the familiar experience of things being "there," no matter how sophisticated and detailed our actual penetration beneath the surface of appearances may be.

Eddington makes us aware of the fact that the problem of how to distinguish truth from mere appearance is embedded in the problem of the nature of the thing. His argument reflects the nature of the transformed way of thinking and experiencing the world that arose at the dawn of modern physical science. Before Kant's Copernican revolution, it was metaphysics that was in foreign territory, dealing as it did with the shadows of what lies beyond experience (the three ideas of reason: the soul's destiny, the universe, God). A reversal has taken place so that it is now metaphysics that is equipped to deal with experience (table A, or in Kant's words: the conditions of possibility of experience are also the conditions of the objects of experience), whereas physics in its post-Newtonian development has come to deal with a world of shadows and symbols (table B). When science has come the closest to its own understanding of the ultimate ground of being, the mediation between the two worlds points to an indeterminate horizon of pure reflection. The history of exact natural science after Newton has slowly unpacked what Kant already says about the absolute totality of things.

Kant's pure reflection results from a combination between two answers concerning the nature of a thing, a combination which is anything but trivial. The skeptic agrees with the scientist: the mystery of substance is solved by doing away with substance altogether. Generalizing from particulars to universals does not presuppose that anything apart from the particulars themselves are given in primary experience. The dogmatist claims that there must be something inherent in the things themselves, as particulars, apart from our experience of them, which is such that we only meet them as particulars; no concept can erase the substance completely. Kant's solution works something like this. Maintaining the dogmatic position without qualification would amount to little more than postulating the very thing we are after, that is, the "what is" inherent in things. Our concepts must include a trace of the unknown substance from which they originate: Kant calls it the transcendental object = x. The very use of an "x" reflects what Eddington said about our being rationally limited in prejudging the question of the ultimate identification of two truths (one for our experience of things, one for the things themselves). The unknowable transcendental object is the only possible worldly counterpart of our own unguided reflection.

Our familiar, everyday experience is one of endless particular things, but scientific understanding is bound sooner or later to connect up with these particulars, even though the intrinsic features of any one particular thing are definitively out of reach. In experience we encounter particular things only: a particular thing here, another there; this one here, that one over there. This is the content of ordinary, familiar experience: nothing but particulars, one after the other. But in scientific experience, or indeed in any judgment in which the understanding is involved, the "this one" which distinguishes one thing from another is simply passed over. This sort of attitude arises from an interest in the connections between various things, that is, in more than just one thing. How shall we understand a particular as particular, since understanding supplies universals? The only possible "middle road" that ultimately links the two worlds together must lie in the *general* characteristic for a thing to be *this one*. What is given in intuition as particular must already contain something which opens it up to the universal. Now, *already within the content of experience* we find things that *appear* to be virtually the same—two tables, two trees, and so on. Kant radicalizes the situation and takes the example of two drops of water (A263–64/B319–20). Two drops look very similar indeed. I can barely distinguish between them. The original problem with one substance is now multiplied to the point that one substance could belong to more than one particular. A distinction must be made between the following two views. (1) The two drops are qualitatively identical; no matter how hard I look I shall never find a feature that distinguishes one from the other. (2) The two drops only appear to be identical, but if I take a sufficiently close look I shall see that they are in fact different.

The middle road is Kant's experiment with concepts. About the two drops, or any two things that seem indistinguishable, I can make either one of the two suppositions. (i) I can suppose that they *are* intrinsically alike (qualitative identity); what still differentiates them is space. The two things that are supposedly intrinsically alike are in different places; and if they were to occupy the same place, they could do so only at different times. What differentiates one thing from another is no longer a matter to be decided by some sufficiently close look. Nor shall we say that their difference is simply numerical: we are interested in the qualities of things as such. (ii) I can suppose that the similarity of the two things is mere appearance *only*, that is, the two things are "in reality" two different things, apart from their being situated in space and time. Here again we may forget about any sufficiently close look that would enable us to distinguish one from the other,but for different reasons. Something "in" the things, beyond all perception, is assumed to be responsible for the differentiation. This something is a prejudgment concerning the intrinsic reality of things. The possibility of such a difference, beyond the scope of our intuition, cannot be simply discarded. The price to pay for this is to ascribe a duplicity to space. It is precisely because things are in space that we cannot know them in their intrinsicality, but nothing can be prejudged about this intrinsicality—not even that it is nothing—because space itself cannot be removed without also destroying the possibility of experience (i.e., the possibility for a thing to appear to us). As Kant puts it in his example of the drops, "the mere fact that they have been intuited simultaneously in different spatial positions is sufficient justification for holding them to be numerically different." Counting takes time, but the spatial intuition of two different things at two different places is simultaneous. That is, every part of space is "outside" every other part, but *not* in the sense of units being separated or separable from one another by the operation of counting. It is just this latter property which puts space (as pure form of human sensibility) on another plane, as precondition of any thinking.

Positions (i) and (ii) cannot thus be distinguished simply as critical and dogmatic. Dogmatism (as in position ii) is possible only as dogmatic *haste*. Kant's argument is directed against Leibniz's conception of space as relational property. The relational view denies any relevance of the space-time background in all our attempts to understand the "this one" of a thing; this background is only a secondary aspect of the thing. From this Leibniz developed his principle of the identity of indiscernibles, which states that two indistinguishable things (by whatever means) cannot be two things but must be the same. No two things can be perfectly alike in the universe. This is a principle of knowledge, which Leibniz grounds in an ontological principle, namely, the principle of sufficient reason. Two indistinguishable things are the same thing because it would be absurd to maintain that God would create two perfectly identical things. That is, even if he *could* do so, such creation would have no purpose. God would act in accordance with purely mechanical deeds if he did so, for the repetition of the same is something purely mechanical. Kant's argument against Leibniz is critical precisely in this respect: Nothing enables us to say once and for all that two things apparently alike must be different "in themselves" just because they are two, unless reference be made to the actual intentions of the Creator, which are totally beyond the limits of experience and knowledge. We know nothing about the purpose of creation, no more than we can tell if two seemingly alike things are alike in reality, even after we have exhausted all means at our disposal to investigate the two things in their relevant respects. Our own haste in getting into the thing is reflected in a God that cannot rest content with the monotonous repetition of creation.

But if dogmatism means haste, the critical position intimates patience. The transcendental object = x is not quite the same thing as the thing in itself (noumenon), precisely because it *could* be the same. The transcendental object is the unknowable x that I come up against with when digging underneath the surface of a thing in order to understand it as an individual. Kant writes (A250) that the "transcendental object cannot be separated from the sensible data": it is the x as fragment of the appearance, a limit directly accessible to familiar experience, since a limit under the surface is never presented in intuition. In the inmost recesses of the individual appearance, the x reminds us that no appearance is given to us outside a manifold of intuition. Where, then, does the unity of our experience of the objects of the appearing world come from? If it were to come from the mind alone, then all that would be needed to account for the unity is what Kant calls the "unity of apperception." That is, all unity of experience could be traced back to the unity already contained in the "I think." On the other hand, if the unity were to come from the objects themselves, then we should postulate that all individuals contain something which is responsible for our encountering them in such or such a way. This something is the noumenon, which is *not* a fragment of the appearance, since it means "a thing so far as it is *not an object of our sensible intuition*." Kant goes on to suggest that the only positive sense of a noumenon is that it could be an object for a sensibility totally alien to ours (an intellectual intuition). This is not an x, because we know at least (as Kant indicates in the teleological part of the *Critique of Judgment*) that, for the higher understanding that would think such an object, this thing could not be other than the way it is thought; its adherence to a manifold is eliminated. Only the transcendental object maintains a relation to thinking as it relates to the manifold of appearances. Kant calls it "the correlate of the unity of apperception."

The thin borderline that separates the transcendental object from the thing in itself is put to work in the second antinomy of pure reason. This antinomy deals with the absolute completeness in the division of a given whole entirely contained in the field of appearance. The thesis claims that every composite substance is made up of simple parts. The proof begins by assuming the opposite proposition, namely, that composite substances are not made up of simple parts, from which a contradiction follows. But this assumption rests essentially on the power of *thought* alone: "If all composition be then removed in thought [*in Gedanken*]" (A434/B462), nothing at all will remain. The argument is not concerned with the physical possibility of the actual division of matter. The parts of matter could be so tightly attached to one another that no finite force could sunder them, in which case we would have only a physically indivisible compound rather than a simple substance. The proof of the thesis discards such a superior force; for otherwise we would fail to appreciate how much *we* constrain nature. Likewise in the antithesis, which asserts that no composite thing is made up of simple parts. Here, Kant argues, "it is not sufficient to find for the pure concept of the composite formed by the understanding the concept of the simple; what has to be found is an *intuition* of the simple for the intuition of the composite" (A441/B469). This proof tends to reduce matter to the space it occupies (in the sense of *einnehmen*, extending through), not the space it fills (which would be in the sense of *erfüllen*) by means of a force. The antithesis is thus quite compatible with an atomic theory of matter. Even if matter turns out to be, in fact, physically divisible (owing to the causal laws of the world, the parts might not be physically separated or separable), the atom remains spatially divisible because it is posited in intuition.

In the thesis, the emphasis on thinking at the expense of familiar experience has certain implications. If all composition were removed in thought, we would face the following alternatives: (1) simple parts are left; or (2) nothing at all is left. Kant refutes (2) because composition can only be a contingent, not a necessary relation of substances. This decision rests on the acceptance of the Transcendental Aesthetic, according to which the relation between parts in space (and time) as a manifold of intuition *is* necessary—which precisely deprives the manifold of any composite character. In the case of space and time, it is thus impossible to remove all composition in thought, since we apprehend the parts only as limitations of the whole, not the whole by a synthesis of its parts. Similarly, Kant dismisses the alteration of the states of substances as a case of composition. Dividing a process of alteration (such as change) into simple parts is *not* tantamount to dividing into two or more substances the thing which the states are being altered into. This agrees with the thesis of the first antinomy, which showed why and how causality gives out at some point as we rise to the absolute completeness of the composition of all appearances. In the case of the division of a given appearance, causality is ruled out from the start. The dogmatist believes nature can be constrained by means of pure thought, independently of the forces governing nature.

In the antithesis, it is the intuition of space which is endowed with a power, namely, the power of transforming the simple into a composite of substances, "since everything real, which occupies a space, contains in itself a manifold of constituents external to one another" (A435/B463). Failing to grant the intuition such a capacity, we would repeat the mistake made by those whom Kant calls the monadists. Instead of presupposing space as a condition for possibility of the objects of experience, we would take "instead these and the dynamical relation of substances as the condition of the possibility of space" (A441/B469). Here, for the first time, a dissymmetry appears in the overall structure of Kant's argument. Unlike the antithesis in the first antinomy, the antithesis in the second is *not* an amplification of the local rule of causality, which insists that a further member in the series of appearances can always be found. Thanks to the rejection of those concepts of force and action that underlie the rule of causality, the same result can now be achieved. The second antinomy explicitly rules out any dynamical relation as a causal factor explaining the origin of space. The nature of space is not accounted for by the dictatorial imposition of certain ends that would be borrowed from the natural experience of the world. The chief lesson of the second antinomy is thus that the power of intuition is indicative of a certain autonomy in sensibility thanks to which it resists complete assimilation with the causal laws of nature. The argument takes exception to the perversion of reversed reason that prevailed in the first antinomy, where the regressive synthesis was acknowledged to result from reason's acquiescence in letting itself be seized by an origin for which it has no corresponding intuition.

Whether this perversion be accepted, as in the first antinomy, or rejected, as in the second antinomy, its ultimate implications are found in the backdrop of the demonstration of the *third* antinomy, which deals directly with the clash between causal determinism and autonomy (transcendental freedom). We shall get around to this point in chapter 7. Our first task is to see what happens to the fundamental opposition, between the familiar world of experience and the world of reasoning, when, leaving the division within a given appearance behind us, we seek to regain our bearings in the realm of the composition of all the appearances that constitute the world. Let us start again by exploring an analogy with physics and astronomy.

The Historical versus the Speculative Background of the First Antinomy

Kant's antinomies emerge from an analogy with the foundational problems related to the rise of mathematical physics in modern times. But since their meaning is ultimately irreducible to any nontranscendental perspective, their conceptual background can be highlighted by reference to issues that Kant may not have been directly acquainted with. This is particularly true for the first antinomy, which cannot be limited to the published Leibniz-Clarke controversy.

Newton and Leibniz had very little to say on the cosmological problem as part of a fully articulated scientific conception. Perhaps Newton would have entirely refrained from considering the issue, had it not been for the Reverend Richard Bentley, who consulted Newton before writing a series of lectures to be delivered in 1692 at St. Martin-in-the-Fields in London. In these lectures, Bentley endeavored to prove the existence of God on the basis of physico-theological arguments. To this end he set out to avail himself of an understanding as accurate as possible of the physical image of the world, the image that Newton defended. In his letters to Bentley, Newton begins by arguing that a materially finite universe embedded in infinite Euclidean space would not be stable due to gravitational effects. Indeed, through the effects of gravitation on the global scale, matter would tend to collapse and refocus to a center,[4] so that no stable physical system such as the solar system could possibly exist. Yet the conception of a materially finite universe tallies all too well with the requirements of natural theology, since an infinite matter would not be really distinguishable from the receptacle which accommodates it, namely, the infinite and eternal space which Newton equates with the Sensorium Dei.[5] Bentley then decides not to leave Newton a leg to stand, since he goes on to suggest that perhaps we have every good reason in physics to believe that a materially infinite universe should be stable. Indeed, the gravitational effects of an infinite matter should cancel each other at each point: the infinite attraction in one direction ought to be countermanded by another infinite attraction in the opposite direction. On this view, would not an infinite chaos be in equilibrium? The argument, to which Bentley himself is not quite committed, serves the purpose of driving Newton to the wall and testing the strength of his whole philosophy of nature.

As a mathematician, Newton replies that one cannot say of infinites that they are equal, but only that they have no relation whatsoever to proportions;[6] hence they are never equal or unequal. To be sure, thanks to mathematical means—what he calls "certain definite restrictions and limitations"—it is possible to determine some differences or proportions between infinites. But this applies only to the infinite divisibility of magnitudes which are themselves finite: in this case, we do not fall into the contradictory comparison of finite amounts as being at the same time different and equal. What, then, can be the reason for not dealing with infinites as equal amounts, so far as their real effects in the physical world are concerned? Are the "restrictions" and "limitations" applicable here again? Newton does not deny that an infinite universe in equilibrium is conceivable on physical, as distinct from mathematical, grounds. But then, as he argues, "if a Body stood in Equilibrio between any two equal and contrary attractive infinite Forces; and if to either of these Forces you add any new finite attracting Force, that new Force, how little soever, will destroy their Equilibrium, and put the Body into the same Motion into which it would put it were those two contrary equal Forces but finite, or even none at all; so that in this Case the two equal infinites by the Addition of a Finite to either of them, become unequal in our ways of Reckoning."

Instead of solving Bentley's problem, Newton apparently shifts the whole question. He acknowledges that infinites are equal in Bentley's sense, but he tries to show that this has no perceptible effect in nature. If they were real, equal infinites would not actually be different from any other finite quantity, or even from zero. Applying mathematics to nature as a concrete, infinite wholeness, the infinite can be dealt with as if it were nothing at all. But what could justify in turn the addition of a new, finite attracting force, once that equilibrium were actually realized? The addition could be nothing more than a purely local motion, such as the motion of a planet revolving around the fixed sun. Or, if applied to a larger set of bodies, the argument would simply reiterate the already-known conclusions concerning the collapse of any finite extent of matter: the infinite universe outside any finite set of bodies in static equilibrium should provide a centrifugal force inside, and that is clearly impossible if infinites are declared to be equal. Yet Newton intends his conclusion to apply to a total collapse of the infinite universe. Even if the motion of a single body is indifferent to the infinites, the addition of a finite does make infinites unequal, and from this Bentley should be convinced that the total equilibrium of the infinites would be destroyed. That is, the addition of a new force, however minute it may be, does affect the total equilibrium but is not affected by it.

This tortuous attempt to come to terms with the infinite universe puts forward the seemingly irreducible distance which crops up between the mathematical and the physical at the level of global experience. If the infinites are equal, as a physical conception requires them to be, there cannot be any kind of motion, not even local motion; but equilibrium is thoroughly and naturally realized. On the other hand, if the infinites are not necessarily equal, following a mathematical conception, motion is possible but there is no equilibrium. More generally, reference to the infinite merely displaces the fundamental problem of finiteness, namely, the problem of how to account for the emergence of a minimum stability *in* the universe in such a way that the stability *of* the universe itself would follow.

So much for the aporias relating to the spatial magnitude of the universe. Leibniz has reflected on the question of temporal magnitude in terms that are significantly not dissimilar to the foregoing aporias of space. According to Leibniz, matter does not include in itself the reason for its existence, since God alone is *causa sui*. God creates from nothing, and therefore any attempt to place the creative act in a moment of time is wrong-headed. Supposing a first beginning of the world cannot be consistent with an empty time before this beginning, for otherwise time would be something absolute;

but as absolute, time is only abstract, that is, the purely ideal order of all possible and real successions. The real world cannot but emerge from divine eternity, which is both independent from and "above" time. In his fifth letter to Clarke, for instance, Leibniz explains that the difference between the duration of created things and God's duration is one of nature, which makes it irreducible to a difference of magnitude.[7] As such, the creative act is non-temporal. We thus remain free to consider it either as "emanating" from God or as the permanent effect of the divine action. In the former case creation is truly "instantaneous," while in the latter case it is akin to some kind of continuous creation. In both cases, however, what arises is the concrete time that belongs to the series of appearances as it becomes conscious of itself as a series. Even though they differ in terms of their principles of explanation, the moments of "ideal" or "concrete" time are homogeneous to each other. Any explanation of the fact that the world was created at such or such a time cannot be rational: whether we look before or after the event of creation, it is simply impossible to either progress or regress toward it.

Following Leibniz, creation and becoming are beyond the mathematical principles formed by the human understanding, whether they are used in their own right or in connection with the concrete properties of the natural world. Beyond mathematics we find the principle of sufficient reason, which alone is fit to "explain" the created beings. Here again, however, two hypotheses present themselves. Either the perfection of becoming is always equal to itself, but of multifarious nature, or becoming accomplishes a perfection which grows endlessly.[8] Under the first hypothesis, the very notion of beginning collapses, since no reason stands out to admit a starting point to the successive series of phenomena; at any time the perfection of the world has always undergone an infinite number of variations. By contrast, under the second hypothesis, the progression into the series of phenomena could be compared, as far as its perfection is concerned, with the series of natural numbers as they increase from unity. But Leibniz is aware of the following objection: "The analysis of necessities, which is that of essences, proceeds *from the posterior by nature to the prior by nature*, and it is in this sense that numbers are analyzed into unities. But in contingents or existents, this analysis . . . proceeds to infinity without ever being reduced to primitive elements." On the basis of the irreducible distance separating essence from existence, Leibniz concludes that the analogy between numbers and instants of time must collapse. Even though the world could have been created so that its total duration is finite, mathematical analysis could still sail up toward the first contingent in accordance with an infinite regress. Ultimately Leibniz seems to favour something like a compromise between the two hypotheses: a growth in perfection from all eternity. Thus the created world would approach asymptotically absolute perfection, without ever making contact with it neither in the past nor in the future. Yet Leibniz is eager to underline the following point: he does "not venture to deny that there may be a first instant." He seeks a demonstration that would combine rigor and reason, but he fails to find one.

Newton and Leibniz's professions of ignorance about the cosmological problem are strikingly similar to each another. Newton's physical theory seems to "distort" the cosmic dimensions of the real, a distortion that contradicts both theory and experience. According to Leibniz, any mathematical analysis distorts the rational account of the real, since a "reality" could come about that has no reason for being. But over and above this declared impotency, Newton and Leibniz also show us how much premodern philosophy is still present in the cosmic representation of modern natural science, as if the revolutionary employment of mathematical constructions in modern physical science had done nothing to change a basic relation to the absolute totality of things.

Aristotle comes to the conclusion which will be that of the moderns, namely, that any full analogy between mathematical and cosmic genesis must be denied. When Aristotle discusses the attempts to construe the original tale of Plato's *Timaeus* metaphorically, that is, such as the references to "before" and "after" can be understood as a way of talking "for the purpose of teaching"[9] (a way that makes matters more distinctly accessible to knowledge), he points out that the analogy with geometrical demonstration is invalid. As he puts it, "in the making of diagrams, all the initial components having been set out, the same comes about simultaneously, but in the demonstrations . . . the same does not come about simultaneously; but this is impossible." In the case of cosmic genesis, "the things assumed earlier and subsequently are contradictory. . . . There must be a genesis involving the separations [of things] in time as well. In the diagrams, nothing is separated in/by time."[10] Thus, if there were anything like a mathematical genesis of the notions germane to mathematics, it could never be deployed to correspond to the movement from the pre-ordered to the ordered cosmos—a movement that Plato could deploy only thanks to the detour of the myth. Aptly enough, Plotinus reminds us that myths too deploy in a time of their own what is simultaneous in the order of being: "myths . . . must separate in time the things of which they tell and set apart from one another many beings which are together."[11] What about the time of these beings themselves, which always come to us already constituted in some way? This is where the powers of human speech and rational explanation seem to give out.

The Platonic account of the world's formation, and how it evolved into its present form, is explicitly considered merely "probable"[12] by Plato himself. The same conception is found in the first scientific cosmogony of modern times. Descartes claims that his vortex theory of planetary motions has to be founded in a model that accounts for how the vortices emerged from chaotic matter and motion. But again, the model is presented as no more than a fable.[13] Not so Kant, who develops the cosmological idea of reason as a logic of illusion, in which no room should be left for merely speculative imagination or probability.

Taking Illusion out of its Hiddenness

In setting out to define the nature and role of a critique of pure reason, Kant explicitly distinguishes between two kinds of ignorance, the accidental and the necessary (A758/B786). Pre-critical reason arises from the simply accidental consciousness of one's ignorance, which in turn pins oneself down to an enquiry concerning things. The enquiry remains dogmatic as long as it is directed without mediation to the things by means of perception alone. Critical reason transmutes the accidental consciousness of my own ignorance into a necessary consciousness. Correlatively, a critical enquiry concerns the function and therefore also the limits of any possible knowledge; the limits *are* the mediation between the knowing subject and the things. These limits (*Schränken*) are not the same as the boundaries (*Grenzen*) that can be known through unmediated perception of things: only critical reason determines the limits of reason itself (*Grenzbestimmung*). Limits are not as mobile as boundaries, since the accidental consciousness of my own ignorance only incited me to further enquiry, while the necessary consciousness absolves me from any such further enquiry. In this way critical reason forms the basis of a science of ignorance (A761/B789). Why should critical enquiry reach a position of no further enquiry? The answer reveals the problematical nature of the object of reason. What limits the field of experience is something unknown to reason, yet "is still a cognition which belongs to [reason]."[14] Something unknown that constitutes a cognition: this takes us at the heart of the undecidability that characterizes cosmological speculation.

The difference between the limits of knowledge and the limits of things themselves is analogous to the difference between science and perception. That this science cannot deal with the unmediated presentification of an object is made clear when the object is the sum of all the possible objects of our knowledge. As we saw in chapter 3, this sum is represented as a plane with its own apparent horizon, much like the earth's surface being viewed as a flat surface. Dogmatic philosophy, unaware of the science of necessary ignorance, would take this apparent horizon as an effect of the merely factual impossibility for experience to overlap with the totality of objects. Our own limited experience could distort the surface in which the totality is given, yet nothing determinate can be known about it—just how much the original totality is distorted remains unknowable. Critical philosophy begins at the point where dogmatism can only lapse into doubt about the steps which reason takes to master what may lie outside the horizon—or on its boundary-line (A760/B788). The skeptic, this burdensome but indispensable companion of the dogmatic, seeks to bring the dogmatic back to self-knowledge, but this is just what will leave the main issue undecided, namely, "what it is that we can and what it is that we cannot know" (A763/B791). The critical philosopher's effort to determine the limits of reason is also the outcome of a self-examination, a return to the capacities for knowledge. But it is a self-examination which communicates with what lies outside the apparent horizon of all possible knowledge. The critical philosopher must become the onlooker of an impossible spectacle.

Indeed, the unsuccessful dogmatic efforts do not simply come up against ordinary facts: they are *Fakta* for the critical philosopher. The *Faktum* of practical reason is not an ordinary fact in the sense of *Tatsache*. It is the consciousness of moral law, that is, the medium through which reason speaks to itself on the occasion of moral duty that transcends the limitations of the material sphere of experience. Aptly enough, Kant calls this superior fact a "celestial voice."[15] The facts displayed by the battlefield of metaphysical systems in the dogmatic era are the imperatives which impel pure reason to find peace within itself, absolving it from further enquiry into (and torment about) the source of all past conflicts. But what is it that an impossible spectacle makes visible? In the case of practical reason and its law of freedom, the power of the *Faktum* is such that imagination is left with nothing to do when reason applies the moral law to sensible objects: the mediation is *entirely* provided by the understanding;[16] if it were not so, the freedom of imagination could always interfere with the categorical givenness of the law itself. In the case of natural laws, reason exerts its regimentation over the imagination so that the latter is not simply left adrift to its own dreams. But imagination is at least capable of *inventing* (A769/B797). Indeed, the concepts of reason have undoubtedly no object in a possible experience, yet they are related to what Kant calls "heuristic fictions" (A771/B799).

How does the inventing capacity of imagination operate in the world of theoretical reason? It would not, Kant argues, be distinguishable from mere opinion, if it were not preceded by something absolutely certain from which it nurtures itself, namely, *the possibility of the object itself*. This reference to the possibility of an object lies at the basis of this new type of mind's self-examination, which does not merely lapse into skepticism. In opposing dogmatic assertions, skepticism is a dogmatism which simply takes them backwards—those assertions that claim to make an unmediated contact with the object. Hence the distinction between skepticism and what Kant calls skeptical method, which is the method proper to the critical enquiry (A423–24/B451–52). This method, Kant emphatically points out, aims at certainty. In the quarrel between dogmatists and skeptics, each side defends its viewpoint "sincerely and competently," but critical reason has no interest in an opinion that falls somewhere between them; critical reason wants all or nothing (A775/B803). Neither the dogmatist nor the skeptic ever infringe on the laws of logical reasoning. Instead, the object of controversy separating them is quite special: a deceptive appearance. No systematic doubt of any kind would ever be strong enough to reject such an appearance out of existence. What is deceptive in the logically flawless reasonings of both dogmatists and skeptics is the way they apply the laws of logic to possible experience: they have their own logic of illusion, which they cannot become aware of.

In the Transcendental Analytic, Kant has distinguished between the two fundamental cognitive faculties: intuition and thought. Thought itself splits up into two faculties, understanding and reason; the concepts of the former are the categories, and ideas are specific concepts pertaining to the latter. What is the difference between these two sets of concepts? Kant argues that the difference is revealed by their relation to the object. Whereas the laws of pure understanding determine the possibility of objects of knowledge inasmuch as these are given in sensible intuition, reason is this dialectical employment of understanding in which "we venture . . . to judge synthetically, to affirm, and to decide regarding objects in general" (A63/B88). This dialectical employment of the understanding is called *hyperphysical*. Whereas from the principles of pure understanding arise the conditions without which an object of knowledge could never be constituted (a logic of truth), reason constitutes objects as deceptive appearances. As a system of pure reason, metaphysics exhibits all of philosophical knowledge, whether this knowledge is true or illusory (A841/B869). Therefore, if there is a transcendental logic of illusion, it cannot be reduced to an anti-dogmatic criticism of dogmatic assertions.

Kant refers to two kinds of deceptive appearances (A297/B353–54). One is within reach of experience: the perception of light rays. A straight stick appears bent to an observer when it is immersed in water. If we were to know appearances only, how could we ever hope to make sense of "true" appearances? A notion of the *object itself* is needed to support the distinction between true and deceptive appearance in this case; this is the object that can be either immersed in water or seen in the open air. Another deceptive appearance is something like "the beginning of the universe," which we cannot manipulate at all. If it is ever to be given, it must be given absolutely as it is. A higher notion of object would thus be needed in order to see whether it is actual—the *object in itself,* the experience of which is not necessary because it is not even possible. Because the object in itself vanishes as a possibility as we try to experiment with it, a judgment concerning this object, as dialectical employment of the understanding, is related to it as "object in general."

Is the object in general the point where object itself and object in itself converge? Within the sphere of experience, the temptation looms up to call on the notion of object in itself as a superior justification for the notion of object itself, since the realm of possible experience must depend in some way on such a thing as the beginning of the universe. The convergence is suggested by the continuity between the two employments of the understanding, empirical and hyperphysical. Indeed, Kant believes that the former is an *extension* of the latter. Extending the principles of possible experience to the possibility of objects in general enables us to assert the objective reality of those concepts for which the objects lie beyond the limits of all possible experience; both the procedure of extension and the asserted objectivity are *transcendent* (A781/B809). Unlike a transcendental principle, the transcendent principle passes beyond the limits of possible experience without falling into an error that would be rectifiable in some way (because it would belong to the faculty of judgment) (A295–96/B352). Because it is incorrigible, the transcendent principle bears a resemblance to an imperative—it embodies the command to transgress the limits of experience. That is why the thing in itself never quite loses its relation to a possible perception, though Kant claims that it can only be thought. That the thing in itself *could* be resolved into an appearance is supported by the obscure attachment of the thing in itself to a perception that is always delayed and never actualized.

The perception of given objects within the limits of sensibility vouchsafes the possible truth of knowledge. Why, then, should a renewed emphasis on perception—the dogmatic obstinacy to ignore a science of ignorance—lead this science of ignorance to a critique of hyperphysicality? Are not the ontological propositions allegedly demonstrated in speculative philosophy and natural theology examples of pure thinking that is *substituted* for an impossible perception, rather than speculations that have become empty because they have blown the field of perception out of proportion? When Kant invokes the inextinguishable desire that animates our reason to transgress limits, he means that if a transcendent principle of explanation would at some point break off from the natural explanation, this would put an end to our enquiries by substituting the utter incomprehensibility of a preconceived principle for an actual comprehension. For example, the concept of what is absolutely primordial would be substituted for the difficult but still practicable explanation of worldly objects (*Weltobjekte*) (A773/B801–2). What is purely preconceived is denied any right to make up for the lack of physical principles of explanation, *even* when the power of the latter might give out. Kant is simply not interested in holding up as a transcendental hypothesis—an opinion that has some degree of probability—a situation in which a natural force might transgress its own laws and find the conditions of its being in a supernatural unconditioned being. In the context of mathematical physics, Galileo had already expressed the same idea when he argued that "nature never transgresses the laws imposed upon her";[17] by analogy with natural science, Kant argues that "no natural force can of itself deviate from its own laws" (A294/B350). But if, as we have seen in Chapter 3, natural science provides a minimal curved surface of the world that metaphysics uses to determine the complete circumference, why are we not allowed to postulate such a transgression as a transcendental hypothesis?

More generally, Kant never really explicitly justifies his theory of necessary illusion, according to which the unobserved influence of sensibility over the understanding is the principle of error. Why is it not the other way around—an influence of some special type exerted by the understanding upon sensibility, such that the otherwise regular determination of sensibility by understanding is disrupted? In the Transcendental Deduction of the categories, Kant shows that no a priori knowledge is available to us, except if it relates to objects of possible experience, but, significantly enough, the categories are not limited *in thought* by the conditions of our sensible intuition. The categories have an unlimited field for themselves when they are taken in isolation from the actual content of experience (B166 note). Why is not this unlimited field responsible for the illusion proper to pure reason, which is the domain "which recognises no limits of demarcation" (A296/B352)? Is not the action of something otherwise passive *less* natural than an uncontrolled release of concatenated mental energy accumulated in the understanding? How could this seemingly unnatural action have the character of necessity that is bestowed upon the transcendental illusion? Introducing such a paradox is not the least of the enigmas of the *Critique*, a paradox that must itself account for a paradoxical knowledge—that with which reason enacts judgments about things that cannot be sensed in any way.

Remember that the antinomies are themselves the result of a transformation of the natural antithetic of pure reason, a transformation whereby reason displays its own laws. Thus, Kant's decision to ascribe illusion to the influence of sensibility over the understanding receives a modicum of justification if we start by looking at some of its implications. The rejection of some dysfunction in the application of categories to sensibility implies that any transcendental sense of categories (which would make them operate within the understanding as their own field) is already a non-sense, and this non-sense is *not* that of the rational ideas. It would be impossible to sever this non-sense at its source if we had a faculty of *originary imagination* producing an object of an entirely "new quality that does not allow of being given in experience" (A770/B798). As an example of such empty figment of the mind, Kant cites the mechanism of an attractive force without contact. Interestingly enough, in his letters to Bentley, Newton had already argued against the force of gravity as an action-at-a-distance that would operate instantaneously in the void without intermediary: "Gravity must be caused by an Agent acting constantly according to certain Laws; but whether this Agent be material or immaterial, I have left to the consideration of my Readers."[18] When Kant points out the need for a material explanation of the nature of gravity (a force acting by contact between material bodies), he reiterates the need to eradicate all traces of skepticism. The theory of gravity may be certain in its theoretical predictions of the motions of heavenly bodies, but, as Newton's own confession of ignorance reveals, this is not a certainty which is rooted in the possibility of the object "force" itself. That is why mere opinion (in the sense outlined at A822/B850) concerning the way it actually operates would still be compatible with it, at least as long as the material explanation has not been fixed metaphysically.

The first effect of Kant's theory of the unobserved influence of sensibility over the understanding is thus to put out of play a faculty of originary imagination that would move freely in the unlimited sphere of pure thinking. To what extent, then, is the unobserved influence dependent upon an "originary reason"? Pure thought cannot be simply blind to its own meaning. We recall that intuitions without concepts are blind to their own meaning, whereas concepts without intuitions are empty. Sensibility is the seat of sense. A bare concept is thus more than blind to its sense: it is completely senseless; it does not even have a meaning that remains withdrawn (A240/B299). Indeed, if the forms of thought happened to lose a reference to sensible intuition, "they [would] have *even less* meaning than the pure sensible forms" through which an object is given; as for the combination of the manifold of intuition in the understanding, it would now operate emptily, and so would signify "*nothing* at all" (B305–6, emphasis added). The function of the concept is to throw light upon the sense concealed within intuition, but by itself the concept is deprived of sense. Kant goes on to argue that conversely, besides the categories, there must exist pure concepts of reason (the ideas) which, when taken independently of sensible conditions, cannot be different from these categories. But the action of sensibility over the understanding has no equivalent in the ideas, because the special property of the ideas is that they are *never* empty. If there is anything like an *originary* unity between thought and intuition, it is exhibited in the ideas and their play of illusions, not in the categories.

Indeed, even if they soar far above the teachings of experience, the ideas continue to move within a horizon of experience that is constitutive of them. When Kant specifies the kind of intuition that is required in order to demonstrate the objective reality of the categories, he indicates that "we need, not merely intuitions, but intuitions that are in all cases *outer intuitions*" (B291). But this outerness is *always naturally* available in the ideas. They seem to have objective reality, because the transcendental appearance never ceases

(A297/B353), even after it has been detected critically in accordance with the innerness of self-examination. Thus, the identity between the category and the idea rests on the fiction that consists in sundering thought from sensibility, but the fiction is itself suggested by the idea inasmuch as it alone never lacks a *visible* reference to outer intuition; the categories only have a reference to outer intuition, which does not become visible until sensible intuition has bestowed a sense upon the forms of thought. If we tried to annihilate the appearances produced by reason, we would fall into the following trap: they would have been annihilated only in accordance with the very illusion of reason that produces them. Thus, the invisible influence of sensibility over reason is the condition of absolute visibility of the objects that result from this influence. Thinking the absence of a transcendental appearance means doubling that very same appearance. This logic of illusion defines the phenomenality of thought, which is more originary than pure thought itself.

The difference between understanding and reason in terms of the horizon of sense attached to each of them initiates a movement of thought which will culminate in Kant's assertion that reason is actually the touchstone of the truth contained in the rules of the understanding: "the hypothetical employment of reason has . . . as its aim the systematic unity of the knowledge of understanding, and this unity is the *criterion of the truth* of its rules" (A647/B675). The pure concepts emanate from the understanding, which alone is capable of producing them (A408–9/B435). But by means of such a production the understanding still fails to open itself up to its own infinite field, and thus does not manifest its freedom, since it is hampered by the "unavoidable limitations of possible experience." Reason does not generate any concept, but it bestows freedom upon the understanding by preserving at all times a *relation* to the empirical, even though the empirical sphere itself may be overstepped. How does this relation get around the alleged necessity of the object of the idea?

Instead of eliminating the transcendental appearance, we could see what happens if the unobserved influence of sensibility were to be controlled in some way and become conscious of itself. Would not the ideas, then, be totally indistinguishable from the categories in their empirical employment? The reference of categories of understanding to the outer intuition that bestows sense upon them becomes visible, because a transcendental deduction of these categories is possible. Thus, the categories "do not represent the conditions under which objects are given in intuition. Objects may . . . appear to us without their being under the necessity of being related to the functions of understanding" (A89/B122). Taken by itself, an appearance does not need the support of the functions of thought: it is free-floating, suspended nowhere, pent in the free play of pure appearing. The aim of the deduction is to eliminate this free play by connecting the appearance with the fixed subjective conditions of knowledge. But the moving field of pure appearing always exceeds the steadiness imposed by the understanding, because it is indifferent to whether or not an appearing appears, whether or not it exists, or whether or not it is merely illusory. The categories regulate the field where the question as to whether an object is something or nothing, is decided: "As the categories are the only concepts which refer to objects in general, the distinguishing of an object, whether it is something or nothing, will proceed according to the order and under the guidance of the categories" (A290/B346). We could say that the appearing of something is an appearance. But the original field of appearing, in which something is not distinguishable from nothing, always exceeds the field of appearances. The ideas remind us of this fact, because they take up again the field of pure appearing at the point where the understanding took leave of it. The objects of the ideas, considered in their relation to the categories, are pure appearances, that is, appearances of *nothing*—whereas the understanding is always concerned with appearances of something = x. Since reason compels the understanding to divide up into an empirical and a transcendental employment, it follows that the necessarily antinomic conflict of reason with itself expresses the gap between something and nothing: the understanding makes nothing of the transcendental employment of its own categories.

A certain necessity is now attached to the fictitious moment that allowed us to identify, in the absence of any sensible condition, the categories with the ideas. Only the understanding in its empirical employment achieves a balance between thought and intuition that does not favor one over the other; when the pure appearing is something, the encounter between thought and intuition is also fully conscious of itself. This balance achieved by the understanding is expressed in *rules*, which are defined as the understanding's power to subsume appearances under a unity. Reason, then, is the power to subsume the rules of the understanding under a unity of appearing and appearance; this unity is achieved by means of *principles*. A rule orders disparate individuals as they fall under a universal, whereas a principle never quite loses sight of the continuity between instances of a universal because it *derives* a particular from a universal. A rule deals with propositions of the sort "X is a case of A," whereas a principle deals with inferences of syllogisms ("X is A" because it is B, and all Bs are As). *Any* mode of intuition pertaining to human sensibility is itself derived (*intuitus derivativus*), not original (B72), simply because we ourselves are dependent beings as far as existence is concerned. As a result, the reasonings of reason are themselves a certain mode of intuition, in the sense that they express what is most eminently human in the way we intuit. The principles are instances of synthetic knowledge by means of concepts, which the understanding alone is unable to provide, since the universal and necessary propositions derivable from it are not possible a priori if they are not extracted from pure intuition (in the case of mathematics) or from possible experience. Reason is concerned with objects indirectly precisely inasmuch as it is concerned directly with what is essentially human in our mode of intuition. Consequently, reason assigns a higher unity to those concepts and judgments which alone result from the combination of thought and actual sensible intuition. Viewed from this perspective, reason is *not* one of the heterogeneous elements of knowledge without which no limit to any science could ever be determinable (A844/B872). Instead, it bestows unity upon the various ways through which the two fundamental heterogeneous elements of any knowledge (intuition and thought) may be combined with one another. It transcends the limits that are set critically to every cognition without ever infringing the laws of transcendental logic.

On the Logical Employment of Reason

This transcendence does not contravene the laws of general logic either. Just as the categories are extracted from the logical functions of all possible judgments, so to the dialectical reasonings must be derivable from that particular species of judgment that determines every possible relation between our representations. To the three kinds of judgment proper to the category of relation, there will thus correspond three kinds of reasoning: categorical, hypothetical, and disjunctive (A304/B360–61). Why should the category of relation become central in the Transcendental Dialectic?

The answer is that the inseparability between thought and sensibility can be expressed only in connection with this category. As the highest principle of all synthetic judgments brought up by the understanding states: "every object stands under the necessary conditions of synthetic unity of the manifold of intuition in a possible experience" (A158/B197). Therefore, the task indicated by the faculty of reason is to seek the highest ground upon which this principle itself rests, the ground that this grounding principle *is*. Kant says that the a priori principles of pure understanding are already doubly named: they ground all other principles, but they are grounded in none themselves (A148/B188). There is no more universal knowledge than these, in the following precise sense: categories are not yet knowledge, only in the subsumption of a representation *under* a category does anything become known. The ground of knowledge lies under, deeper than the categories. Kant believes he needs this firm ground in order to bring home his main point: that the traditionally highest principle of logic, noncontradiction, is limited to analytic judgments. No analytic judgment can solve the problem of the possibility of synthesis, which requires that we "go out beyond the given concept in order to consider something wholly other than what was thought in it, to be in relation with it" (A154/B194, translation modified). What is found under is wholly other. Synthesis, even that accomplished by Kant's mathematical and dynamical principles, involves relation, a holding together in one, a synthetic unity that nevertheless is continually accomplished by a holding together. Moreover, such a relational holding together is with something wholly other, beyond the given concept, and beyond where we are too, at least with our concept, for we must go out to it.

At first, it seems that this wholly other could be any other representation, which I must combine with the given one. But I do not need a priori synthesis for connecting two empirical concepts, say, "brown" with "table," because the concept of "table" does not contain "brown," and vice versa. I do not need synthesis even to know, strictly speaking, that cause and effect go together; this conjunction is analytic, as much as that "husband" can be derived from "wife" or "no wife" from "bachelor." That causality structures appearances in experience, and under what conditions and to what extent it does so, for all these I need a priori synthesis: the relation to what lies under appearances as their condition has to be extracted. Then, in the Dialectic, where Kant examines the propositions that go beyond the given manifold of intuition, the three ideas of reason are derived from the *category* of relation: that is, the relational holding together is now with something wholly other that is itself beyond the given manifold, not just the given concept. This explains also why, in its logical employment (A301–2/B358), reason seeks the general condition of its judgment, that is, the unconditioned which allows the unity proper to any conditioned knowledge to be achieved.

Reason is initially described as "elaborating the matter of intuition and bringing it under the highest unity of thought" (A298/B355). In this respect we can see that the activity of reason bears at least a similar relation to its objects as do the faculties of sensibility and understanding. Each is given a manifold which it unifies: sensibility synopsizes the a priori forms of intuition, imagination synthesizes the manifold of intuition, and the understanding unifies this synthesis through original apperception (A94). The definition of reason will consist in indicating what diversity it shall perform its operation of unification on, and the question of reason will concern the form and significance of this operation. This form and significance mark the division of reason into its two modes of employment, logical and real; but, given this division, "we are constrained to seek for a higher concept of the source of this knowledge" (A299/B356).

The major premise of a syllogism, which can be any universal proposition, is a principle with respect to that which it subsumes, but this does not mean that the conclusion is necessarily knowledge from principles. The major premise could, for example, be derived from experience, through induction, in which case the conclusion of the syllogism would be empirical knowledge, determined a posteriori. Even general mathematical propositions cannot be said to produce knowledge from principles. I may say that there can be but one straight line between two points, but if I know this to be true it is not because "I apprehend this property of straight lines in general and in itself, from principles, I apprehend it only in pure intuition." And just as mathematical propositions rest on pure intuition, the propositions of pure understanding (e.g., that everything which happens has a cause) rest on possible experience. But reason rests on concepts alone. As a consequence, the diversity to which reason will supply unity is the manifold knowledge of the understanding, which is extraordinarily versatile: this knowledge may involve appeals to pure intuition, empirical intuition, or even possible experience, but reason can also be seen in operation when divorced completely from the possibility of experience. What, then, is the character of unity which reason provides, and the modes whereby it does so?

With respect to its logical employment, the form of reason's activity can be explicated along the following lines. Reason, Kant says, "has long since been defined as the faculty of making mediate inferences." Some inferences are immediate, but since the process of making inferences comes so naturally to us, we tend to lose sight of the distinction between a mediate and an immediate inference. Immediate inferences are inferences of the understanding (*Verstandesschlusse*), whereas mediate inferences are inferences of reason (*Vernunftschlusse*). What is the difference? What form does mediation take in reason? Kant writes: "If the inferred judgment is already contained in the earlier judgment that it may be derived from it without the mediation of a third representation, the inference is called immediate" (A303/B360). The manner of containment which Kant refers to in connection with such reasoning is strikingly reminiscent of the definition of an analytic judgment, according to which the predicate is already contained in the subject. Here, Kant refers explicitly to a difference in form, but not in matter. When one concludes "some men are mortal" from the premise "all men are mortal," one has made no change in the matter of the original judgment—subject and predicate remain the same, they have simply been placed into a new relationship. A change in form has taken place, according to a certain rule. The totality of such rules are none other than the logical functions of judging which formed the basis of the table of categories in the Transcendental Analytic. What makes the inferences of the understanding immediate is that they occur in one step; the *Schlussfolge* which connects premise to conclusion is given as part of the premise and temporally determined by its relation to the categories of pure understanding. The temporal determination does not appear in the reasoning, it remains concealed in the judgment itself. In section

44 of the *Logic,* Kant remarks that one can also make a *judicium intermedium* in conclusions of the understanding, but then this mediating judgment is merely tautological, for example, in the immediate conclusion "All men are mortal; some men are men; therefore some men are mortal," the middle concept is a tautological proposition. What this means is that reason need not provide any unity to this operation of the understanding: it is already one. Problems arise when a midpoint is introduced along this *Schlussfolge.*

The addition of one intermediary step effects a qualitative change to an inference. If immediate inferences are characterized by changing only the form of judgment, leaving their matter intact, we can expect mediate inferences to involve a change of matter. Indeed that is precisely what the minor premise does relative to the major premise, that is, it introduces something new, either a new subject or a new predicate. The different modes of mediate inference will, then, be determined by the different possible manners of relating some new subject or predicate to a general rule. What different kinds of syllogisms are there? Kant writes in section 60 of the *Logic*: "Syllogisms can be divided neither as to *quantity,* for every major is a rule, thus something in general; nor in respect of *quality,* for it is equally valid whether the conclusion is affirmative or negative; nor finally with regard to *modality,* for the conclusion is always accompanied by the consciousness of necessity and consequently has the dignity of an apodeictic proposition. It is solely *relation* which thus remains as the only possible ground of dividing syllogisms." In this we start to see how the Transcendental Dialectic is shaped up. The Dialectic does not constitute a particular type of logic, but only the employment of formal logic as an *organon* (the means needed in order to acquire knowledge). (At A61/B85–86, Kant shows that general logic, if viewed as an organon, is always a logic of illusion.) Thus, the reasonings of the Dialectic will not presuppose anything but the principle of hypothetical syllogism (according to which the conditioned is always given through a condition), which lies at the basis of formal logic. The understanding unifies the synthesis of the manifold of intuition (carried out by the imagination) through original apperception. The correlate of this unity of apperception, the transcendental object = *x*, then formed the basis for the elaboration of the categories of pure understanding, insofar as these can be thought of as what can be predicated of the transcendental object. But in the case of reason, the categories of relation will be predicated not of the transcendental object, but of the unity characteristic of the givenness of mediate logical inference as apodeictically certain. With respect, once again, to the logical employment of reason, Kant writes: "the principle peculiar to reason . . . is to find for the conditioned knowledge obtained through the understanding the unconditioned whereby it is brought to completion" (A307/B364). We can say that reason synopsizes the potentially infinite series of pro- and episyllogisms, giving the totality of conditions, which is itself unconditioned because it collapses into the expressed generality of the major premise. But when we move back to pure reason, in its transcendental employment, we once again problematize the unconditioned. The activity of reason becomes synthesis, rather than synopsis, and the unity of the synthesis, the unconditioned, is of a character that it cannot necessarily be given in experience. So, where, in the Dialectic, the elaboration of the categories of relation in connection with the transcendental object issued up the "Analogies of Experience," these same categories taken in relation to the unconditioned will give forth something quite different, trans-experiential analogies.

The Transcendental Amplification of the World

What is a trans-experiential analogy in the particular case of the hypothetical syllogism, which corresponds to the category of causality? What specific form does the correlative problem of the idea of reason have, when its object is the whole universe? The concept of the universe is an idea based on unconditioned totality, but it seeks the empirical synthesis corresponding to this totality. This concept is problematical, not intellectual, because the two following problems are *not* identical: (1) Is there an idea that corresponds to a possible empirical synthesis—an idea that is not too large for such synthesis? (2) Is there an empirical synthesis that corresponds to a given idea, that is, a synthesis that is not too small for such an idea? Bringing the idea down to an empirical synthesis is not symmetrical to raising the synthesis to the idea. The idea of reason is problematical because reason itself is unable to decide which way of questioning is appropriate to its fulfillment. There is no solution to these two problems in which the weights shared by thought and experience could be equally distributed. But the unity brought about by the understanding remains the model for reason. Therefore, in the apagogic mode of proof, pure reason always falls into this paradox that consists in taking possession, for each solution, of that which belongs essentially to the problem underlying the other. Hence a third question: Can an idea and an empirical synthesis accord with each other in such a way as to *not* modify the already acquired empirical applicability of the pure principles of understanding?

The cosmological problem would have a unique solution if pure reason could generate its own concept, relinquish the desire for symmetry between thought and experience which is borrowed from the understanding, and yet not modify the sense of the rules proper to the understanding. For then the agreement between the idea and the empirical synthesis would be achievable—the universe would be an essentially empirical problem awaiting solution from the progress of perception. To be sure, emancipating the already existing concept of understanding from its sensible limitation would lead to the release of the unlimited field of pure thought. But why is there no new concept corresponding to this infinite field? The answer is that pure thought is itself horizonal because it depends on the original phenomenality of thought. What restricts pure thought (sensible experience) is also what transcends it, in accordance with the unsurmountable heterogeneity of thought and experience. On the one hand, if thought dwells in worldly being outside itself, it goes beyond itself; if thought allows the realm outside to come to it, then it understands things regardless of its own efforts, and thus also outside of that which gives them a reason for being. On the other hand, if it eliminates any outside, striving to embrace all into its own infinity, it would continue to produce an outside despite itself, without becoming aware of it—an appearance of an outside (transcendental appearance). In this double movement carried out by pure thinking, in which a realm beyond is simultaneously eliminated and created, the natural antithetic of reason is enacted.

Because of the lack of symmetrical balance between thought and experience in reason's enquiry of the absolute totality, a natural selection takes place among the categories of understanding. Among all categories, "only those in which the synthesis constitutes a *series* of conditions subordinated to . . . one another" (A409/B436) can be interpreted rationally. In order to avoid the division of the idea of world into two ideas, when we isolate a given conditioned and try to find its conditions, the conditions must be regarded as being given together with it at the same time. Thus, the absolute totality required by reason is required only for the ascending series of necessary conditions for a given conditioned: this is the *regressive synthesis*, which identifies the absolute totality with a sum total of past conditions. With respect to consequences, it is the conditioned itself, the conditions of which we are seeking, that would play the role of a condition. But the movement to consequences is, according to Kant, an arbitrary problem that cannot possibly lie within the interest of reason; the form of totality resulting from progression would be "gratuitous and unnecessary" (A411/B438) because it results from the confusing exchange of the conditioned for the condition. Heidegger argues that, at this point, Kant seems to dissociate nature from history.[19] The essential character of a historical event is that it can be understood only from its consequences. Does Kant's argument lend support to the view that historical time is completely arbitrary? We shall return to this issue in chapter 6. For the time being, Kant's treatment of natural causality must be followed to its ultimate implications. The given conditioned is untouchable, it must slip away from the conditioning power of thought if thought is ever to represent its ultimate condition. This requirement is indispensable for laying down claims of necessity and universality at the level of totality.

Indeed, reason's effort to keep the workings of understanding as its model leads it to mimic the understanding's own claims to necessary and universal conditions, even though its object (the absolute totality of appearances) is particular and contingent. Consider the transcendental mechanism by which something necessary and universal must be "added" (in the sense of A1–2) to experience in order to make it possible. How can we assert that such a mechanism is necessary, how can we justify its legitimacy with regard to experience which we always depend on and with which all knowledge begins? If reason, as we have seen above, is to possess the ultimate touchstone of the truth of the rules of understanding, why could we not argue that their a priori character is itself not dissimilar to a transcendental illusion? This would be an illusion through which our cognitive faculties are urged to seek in themselves what is necessary and universal, thereby depriving themselves of any means to understand nature as it is in itself. Kant's strategy to eliminate this undesirable implication of his own skeptical method is as follows. The question of deciding whether a priori knowledge is itself necessary is *not* a question to be solved a priori, in accordance with norms of necessity and universality that reason could bring in. Instead, the question does not concern concepts, but their object. The possibility for knowledge to be constituted a priori finds its own condition of possibility in the general problem of reason: what is the totality of conditioned *appearances,* which is so absolute that *nothing can be added* to the experience we have of these appearances in order to understand them in their necessity? In Kant's formulation: "the idea of absolute totality concerns only the exposition of *appearances*, and does not therefore refer to the pure concept, such as the understanding may form, of a totality of *things in general*" (A416/B443). Suppose the understanding did represent the totality, apart from sensible conditions; in that case, "we can indeed *at once* assert that for a given conditioned, the whole series of conditions subordinated to each other is likewise given" (A416/B444, emphasis added). But the relation of reason to appearances introduces a "special limitation" by which the immediacy claimed by the understanding is lost. The problem of necessity—the search for the unconditioned—is thus displaced or postponed. Reason is compelled to content itself with a makeshift when it "adopts the method of starting from the idea of totality," because "what it really has in view is the *unconditioned*, whether of the entire series or of a part of it."

This move which lowers the claims of reason from unconditionality to totality is introduced by Kant as follows: "Since, then, the *unconditioned* is necessarily contained in the absolute totality of the regressive synthesis of the manifold in the field of appearance—the synthesis being executed in accordance with those categories which represent appearance as a series of conditions to a given conditioned—reason here adopts . . . " The German reads: "Da, in der absoluten Totalität der regressiven Synthesis des Mannigfaltigkeiten in der Erscheinungen . . . " Thus, the regressive synthesis does not bring forth a modification of appearances, as if the synthesis would destroy them and retain the manifold that they contain. To the contrary, the regressive synthesis *constitutes* an appearance in accordance with the categories. The German continues: "nach Anleitung der Kategorien, die sie als eine Reihe von Bedingungen . . . " The "sie" is correctly identified in the translation as the appearance, not the regressive synthesis. Only thus can the constitution of appearances by the regressive synthesis be explained; this constitution itself explains that the unconditioned is necessarily contained in the totality of the regressive synthesis.

The postponing of reason's apprehension of its actual object—the unconditioned—is also justified by another asymmetry in the synthesis toward the absolute totality. For if time is the formal condition of any series, so that antecedents and consequents may be distinguished a priori in accordance with reason's requirements, space is an aggregate, not a series. Whereas any moment of time is given through the totality of given conditions in the past, the manifold parts of space are coordinated rather than subordinated to one another. However, unlike the mode of givenness, the manner in which we apprehend space is still by means of a synthesis of its parts, and this synthesis takes place in time, and thus also in accordance with a series: "Inasmuch as one part of space is not given through the others but only limited by them, we must consider each space, in so far as it is limited, as being also conditioned, in that it presupposes another space as the condition of its limits" (A413/B440). The successive synthesis can be applied to each part of space, with the consequence that each limited part can now be seen as conditioned by another. Space is a series with respect to its limitedness, not its givenness. On the other hand, regression and progression seem indistinguishable for this series, since nothing prevents condition and conditioned from being exchanged; the part of space which is the condition of another part could equally be conditioned by it. Space can be freely cut up into parts, or parts of it can be added in thought; and as limited, any point of space can be used as a possible fixed point (absolute center) from which a series of successive conditions hang together. The confusion between the condition and the conditioned, which violated the absolute givenness of the present time, is now harmless: "the side of the conditions is not in itself distinct from that of the conditioned."

The two reasons that justify the substitution of totality for the unconditioned are thus the following: (i) the asymmetry between the two possible connections between the idea and its empirical synthesis; (ii) the asymmetry between time and space regarding the successive synthesis. No such substitution would be needed if a concept were to correspond to the infinite field of pure thought. When the pure concepts of understanding are liberated from the boundaries of possible experience, the phenomenality of thought has been ignored. Under these conditions, the transcendental unity of apperception would be the subjective condition at the basis of the idea of a complete totality and unity of nature. The understanding needs the transcendental object $= x$ in order to recognize empirical objects under certain concepts: an object is known because *all* objects tend to exhibit a convergence toward this x, which is a common but unknowable object above all of them. The transcendental object can thus be defined as an object *corresponding* to knowledge, and yet *distinct* from it (A104). Likewise, if reason creates for itself nonsensible objects as though they were sensible, all it does is recreate a distance (as norm of all objectivity) between the known and the transcendental with respect to the object, a distance that was immediately jumped over in the act of thinking nonsensible objects. The relations of ideas to their own objects are grounded in a reconstitution of a distance which, from the transcendental viewpoint, can only be *originary*. Hence the illusion of constitutive ideas, which results from the amplification and delaying of truth. Is there an objective basis for this process?

The intuition of external objects, in space and time, represents objects as they appear to us (*erscheinen*), which does not mean that these objects are mere illusory appearances (*Scheinen*). A *Schein* is an *Erscheinung* that has not been phenomenalized by thought, which makes it also independent of our mode of intuiting. But there are degrees of independence. At the limit of separation, there is the noumenon which as an object has lost every determined relation to the subject. On the other hand, when I say that "bodies merely *seem* (*scheinen*) to be outside me" (B69), I am mistaken; but an intuitable relation to the object is nevertheless established. What transcendental ideas seek to preserve is a determined relation of objects to subjective conditions. An appearance is transcendental (*transzendentalen Schein*) when the predicates attributed to the object (by means of the regressive synthesis) concern the absolute totality of objects. Thus, the transcendental appearance is nothing other than the phenomenalization of the transcendental object $= x$; viewed from the categories, it must always miss that which it aims at.

If transcendental illusion is delayed truth, perhaps the *beginning* of the absolutely completed synthesis might conceal a clue to its possibility. Kant writes: "we cannot know, at least at the start of this enquiry, whether such a synthesis is possible in the case of appearance" (A416/B444). Yet, for any knowledge to be transcendental, whether veridical or illusory, an anticipation of the synthesis must be possible. All synthesis takes place in the imagination, which synthesizes the manifold of intuition by means of transcendental schemata. However, no such synthesis is ever completed, because the schema only gives the sketch of a complete object in perception. Imagination, as we noted in chapter 2, is the ability to represent an object in intuition, even in its absence, provided that the object has been present at least once in the past. Thus, if the absolute completeness of appearances were itself possible in appearance, no transcendental schematism would be needed in order to synthesize this manifold. Perhaps, then, the incomplete synthesis of imagination is still at work, which is why the absolute completeness turns out to be impossible in appearance. According to Kant, however, it is the other way around. Imagination is reduced to silence by the idea, which is always at one with itself, not affected by a limitation or an internal division, because it is "independently alike of the possibility or of the impossibility of our connecting with it any adequate empirical concepts."

If the complete synthesis depended on transcendental imagination, the idea of absolute completeness would require an impossible feat on the part of the imagination, namely, an absolute synthesis of conditions. The absolute synthesis of the manifold would produce an empty intuition, that is, an intuition without the presence of an object (as appearance). But such an empty intuition without object is one of Kant's concepts of nothing: "the mere form of intuition, without substance, is in itself no object, but the merely formal condition of an object (as appearance), as pure space and pure time (*ens imaginarium*). These are indeed something, as forms of intuition, but are not themselves objects which are intuited" (A291/B347). Whether it can be satisfied or not, reason's demand for absolute completeness passes the test of imagination, thanks to the subterfuge which consists in setting the pure a priori forms of an appearance apart from the sensible intuition itself.

This "subreption," as Kant calls it, the substitution of subjective for objective conditions, accounts for the claim that rational ideas "follow the direction" (*nach Anleitung*) provided by the categories. Space and time would be concepts if their parts were partial representations of universal space and time. But this is not the case of space and time as pure a priori forms: space, for instance, is a concept only in the very special sense that "it can be thought as containing an infinite number of representations *within* itself" (B40). Space is a thinking intuition, it is thought like an intuition intuits. Kant comes closest to characterizing this duplicity of thought and intuition when, concerning the special kind of unity inherent in space and time, he writes that "space and time are represented a priori not merely as *forms* of sensible intuition, but as themselves *intuitions* which contain a manifold." That is, space and time can also be taken as objects of intuition, not just as forms of any possible intuition, which is just what is required for them to be measurable. Their unity, therefore, "precedes any concept, although . . . it presupposes a synthesis which does not belong to the senses but through which all concepts of space and time first become possible" (B160). (In the first edition of the *Critique*, at A25, Kant spoke of space as being represented as infinite, because if it were not for the "limitlessness [*Grenzenlosigkeit*] in the progress of intuition" that goes immediately beyond every spatial relation, it would not be possible for a concept of spatial relations to sustain the paradox of including in itself the principle of its very infinity.) That by which space and time are *given* as intuitions presupposes a synthesis which can be neither quite sensible nor quite categorial. Thus, it is only in what precedes any limitation (or determination by the concept) that *thinking* space and time as containing within themselves an infinite manifold of intuitions becomes possible.

In the impossible absolute synthesis of a manifold, the forms of intuition are invested with the function of the concepts of understanding. The regressive synthesis in pure intuition constitutes the series of appearances in figure 1.

condition (B) ← given conditioned (A)(= appearance)

↓

given conditioned (C)

condition (D) ←┘

(figure 1)

In imagination an object can be there in representation, even though it is not given. But in the regressive synthesis, the problem is inverse: how can a conditioned be given, even though it is not there? In order for a nonintuitable object to pass off as an object of intuition, that in which all intuition is given (space and time as pure a priori forms of sensibility) must be a spontaneity equivalent to that of thought. The amalgamation of A and B occurs by means of pure forms of intuition only, without transcendental imagination. The present time is the absolute center from which reason starts. First, the extraction of B from A is a temporal synthesis, which liberates an appearance from the weight of its own past that made it a given. In Kant's terms: "in conformity with the idea of reason past time, as condition of the *given* moment, is necessarily *thought* as being given in its entirety" (A412/B439, emphasis added). Then the combined object A + B, in which the distinction between condition and conditioned does not matter any more (no absolute center), is regarded as a given conditioned C by means of a spatial synthesis. Time separates what is united; space unites what is separated. By means of the double action of space and time, an absolute center (human time) remains at all times the condition of the dissolution of all centers (cosmic space); the appearances are constituted by the regressive synthesis so that they are also "given" by it. The transcendental illusion lies in this constitution of appearances as *appearances of appearances*.

No pure thought can relate to itself only; it *has* to relate to an appearance, whether or not an appearance is given. This enables Kant to specify the relation that holds the unconditioned together with the totality: "this *unconditioned* is always contained in the *absolute totality of the series* as represented in imagination" (A416/B444). Indeed, the schema of the category of quantity is the production of time in the successive apprehensions of an object. But in the case of the idea, in the absence of an object to be apprehended as straight appearance, when schematization comes to nothing, the regressive synthesis entirely monopolizes the ability to produce time. Thus, against imagination and its alternations of presence and nonpresence of the object, the regressive synthesis seeks to retrieve an *uninterrupted* union of object and thought.

The transcendental appearance, which is the ultimate object aimed at in this synthesis, thus plays the part of what has become known as the *world-horizon* in terms of contemporary phenomenology. According to Husserl, the all-inclusive open horizon of experience is inseparable from the experience of objects, but it provides an unnoticed restriction (an apparently unchangeable "style" for every given experience) because it is the *pre*condition for ever continuable experiences in which whatever experienceable emerges. As a result, if this horizon is to become itself thematic, that is, if we are to become aware of this restriction, the faculty of imagination is needed in order to make it intelligible: any actuality must now be treated as a possibility among pure possibilities; we are then driven into Husserl's "pure fantasy-world" without which the eidos (unconditioned) of pure factuality could never be brought to light for its own sake.[20] By contrast, Kant's unconditioned is not the always-already-there that has to be clarified by becoming thematic in imagination. The always-already-there is the transcendental object = *x*, which can never become thematic because transcendental schematism sensibilizes our categories *without* providing them with any image whatsoever—not even an image of the *x*. But the duplication of appearance (i.e., the constitution of cosmic appearance as appearance of appearance) takes time. This is not the time of representation, which is always affected by the immediate interplay of presence and absence in the object. Rather, in the transition to cosmic appearance, the time of the regressive synthesis is closest to the enduring time of our own life. Because the synthesis keeps moving off from the immediately given object, only to draw nearer to an absolute origin, this is a time of lesser or greater density, not straight phases of presence or absence, that cannot be represented at all.

In the later, postcritical *Opus Postumum*, reason's search for uninterrupted union of object and thought is radicalized in terms of a reversal in the connection between forms of sensibility and appearances. If space and time are taken to be appearances (objects of the senses), instead of appearances being given in space and time, then the subjective basis of the regressive synthesis can itself appear in some way. It appears as an "indirect appearance," or, as Kant says again explicitly, as an "appearance of appearance."[21] In the

Critique, Kant still works with the distinction between an object that would be given absolutely (to be determined by reason) and the "*object in the idea*" (A670/B698). In the latter case, "no object, not even a hypothetical one, is directly given, [in the sense that it] only enables us to represent to ourselves other objects in an indirect manner." What the postcritical work adds is that the amplification of the givenness of the appearance can be traced back to the subjective forms of sensibility themselves. But in order to represent the forms in this way, a mediation is needed, which is space and time as appearances. Only natural science, which connects space and time to the *motions* of objects in it, is thus able to effect the reversal. Motions of natural bodies are the indirect appearance of the motion of the transcendental idea as it strives to make contact with absolute completeness.

However, the asymmetry between space and time in the regressive synthesis puts severe constraints on the representability of its subjective condition. Any motion of a movable body is always given as dislocated, since it is only relative to a system of reference that can be picked arbitrarily. But the relative motion of the movable body affects the system of reference itself, which must always be capable of being thought in motion relative to a more extended reference frame in which it moves. This regression toward an ever more encompassing space at absolute rest cannot admit of any limit, since all spaces given in experience are relative to an enlarged space which has the peculiarity of being "still material" but "only in thought." Therefore, absolute space in Newtonian physics (in which any point may be used as fixed point) may be defined as an idea of reason.[22] In the motions of the natural world, there is no place for an absolute center. Because this world is universally dislocated, the absolute starting point of a temporal synthesis has no rational counterpart. Motion is a spatialized, not a spatiotemporal phenomenon. One of the fundamental implications of transcendental philosophy for natural science is that, unlike absolute space, absolute time is not an idea of reason.

From the Natural to the Transcendental Antithetic: The Breakdown of Mathematics

While the natural antithetic enables us to identify the unsurmountable problem of pure reason, a transcendental antithetic is defined as "an enquiry into the antinomy of pure reason, its causes and outcome." It deals with "the conflict of the doctrines of *seemingly* dogmatic knowledge" (A420/B448, emphasis added). No rupture with the critical sphere occurs, because the assertions to be examined are not outrightly dogmatic. A natural problem is converted into a lawful conflict. Each of these seemingly dogmatic knowledges, Kant declares, is free of contradiction, so that the conflict is the opposition of these knowledges themselves.

Hegel has pointed out that Kant is inconsistent when he argues that an antinomy is an apparent contradiction of reason with itself (an "unavoidable illusion," *ein unvermeidlicher Schein*), which is the foundation of a "natural dialectic" (CJ 386).[23] Kant simultaneously states that the contradiction is illusory, that is, that it rests on a sophism, and that it is inevitable. But if the contradiction rests on a mere sophism, one would expect it to be avoidable. We need not, however, be surprised about this. The combination of contrivance and naturalness is the essential character of the Copernican revolution in metaphysics. In the transcendental antithetic this combination is pushed to its limits.

The cause of the natural dialectic is the double attachment of reason, both natural and contrived, to the model of allegedly absolute certainty provided by mathematics. When the categorial constitution of experience is transposed from thought to the pure forms of sensibility, the resulting transcendental propositions go beyond the field of sensible experience. Is the transposition equivalent to a construction of the concepts of reason in pure intuition, like that in mathematics? Kant warns that "their abstract synthesis can never be given in any a priori intuition, and they are so constituted that what is erroneous in them can never be detected by means of any experience" (A425/B453). Transcendental propositions would be misconstrued if their syntheses were regarded as mathematical or even mathematizable; reason's appropriation of the field of a priori intuition does not mean that it sets about to construct its concepts in this field. Moreover, it is the insidious inclination toward the mathematization of the natural antithetic which is responsible for catapulting reason into an antinomy.

Mathematics and cosmology cannot deal in the same way with a quantity that presents itself as a whole. In the case of the (mathematical) principle of pure understanding according to which all intuitions are extensive magnitudes ("Axioms of Intuition"), "all appearances are . . . intuited as aggregates, as complexes of previously given parts" (A163/B204). These magnitudes, as we saw in chapter 2, result from the marks that they leave behind them, as past traces, at the time of their construction (as when I represent a line by drawing it in thought). Now, in the case of the cosmological synthesis, the previously given parts could not have been constructed by the mind itself, because this synthesis deals with given appearances. The proofs of both the thesis and the antithesis exhibit the impossibility of constructing, but this is an impossibility that reason still wants to make up for in some way. Thus, in the thesis, a beginning of the world results from the fact "that it is impossible for an infinite world-series to have passed away" (A426/B454), that is, such a series has to be rejected because it can never leave a trace in the present. Of course, the mathematician will not find it difficult to dismiss the transcendental argument: think of the infinite series of negative numbers ending with −1. Similarly, in the antithesis, an empty time that precedes the coming to be of the world is just as incapable of marking its presence in such a way as to leave behind what Kant calls a "distinguishing condition." Only the clash between the thesis and the antithesis gives a sense of why the transcendantal appearance is a necessary fiction. Each proposition is based on an attempt to suppress an impossible construction, an attempt that fails precisely because any actual construction used as a replacement would be merely mathematical, that is, another presentation in intuition.

In order to do geometry, and mathematics in general, the representation of space as an object is needed (B160-61, note). But from the transcendental viewpoint, space cannot be outwardly intuited. The possibility of mathematics depends on the answer to the following question: What is it that is required to represent as an object something that can never be objectified? To be sure, the geometrician will accept that all constructible spaces can be thought as parts of the originally infinite space given in sensibility. However, only transcendental philosophy can *prove* this in terms of what makes a concept constructible. Indeed, the representation of space as an object contains more than the mere form of intuition. It also contains the combination (*Zusammenfassung*) of the given manifold in an intuitive representation, which is provided by the "formal intuition" (*formelle Anschauung*). Whereas the form of intuition *has unity* and gives the manifold of representation, the formal intuition *gives unity* of representation. Thus, what is distinctive to mathematical cognition is that the unity of formal intuition is prior to the concept of understanding, and yet it provides a synthesis that does not belong completely to the senses either. The possibility of mathematics shows that pure sensibility is not pure receptivity: in order for the parts of space to be generated (constructed) in pure intuition, they have to be detachable from it so that they can *also* appear as products of the construction; the detachability is secured by formal intuition.[24] The junction between mathematical science and pure understanding occurs only *after* the figure intuited as extensive magnitude has become independent (as outer appearance) of my capacity to construct it (A163/B203-4). Now, if reason, unlike the understanding, were to imitate mathematical construction, it would have to forge a similar representation of space as an object. But in the proof of the antithesis of the first antinomy, Kant conflates deliberately, in the definition of space that suits reason, what has just been carefully distinguished: the form of intuition and formal intuition (B457, note). Yet one would have expected that he would stick with this distinction precisely at this point, when he wants to prove that empirical intuition is *not* a composite of two separate things—appearances and space, or perception and empty space.

What is wrong with substituting mathematical construction for transcendental appearance will not be fully intelligible until we look at the application of the category of quality to synthetic judgments ("Anticipations of Perception"). Any reality of some degree = X (as intensive magnitude) can be apprehended in empirical consciousness in such a way that this degree can be diminished continuously down to zero. This means that "a synthesis in the process of generating the magnitude of a sensation from its beginning in the pure intuition = 0, up to any required magnitude" (B208) is possible. In this synthesis, the degrees from 0 to X are the unconscious moments of perception. The understanding acknowledges the unknowable part of its attachment to the world by filling the gap that separates the beginning of reality from the beginning of the perception of this reality. But reason is interested in the uninterrupted connection between experience and thought. Therefore, instead of deriving it, it starts from the continuity itself, which implies that reality be amplified *beyond* the "required magnitude." But how can reality be amplified without changing anything to the actual degree of a given sensation? The unknowable part of our attachment to the world must now be given another expression. The answer is that any degree X is itself generated by a nonzero *quantity* in pure intuition. This is the paradox of pure reason, since, of course, this quantity cannot correspond to the category of quantity. But herein lies reason's final aim, which is to bring about a unity between categories.

What is such a quantity? According to Kant, it is possible to *intuit* an *indeterminate* quantum as a whole. In a note to the proof of the thesis in the first conflict of transcendental ideas, he writes that "an indeterminate quantum can be intuited as a whole when it is such that though enclosed within limits we do not require to construct its totality through . . . the successive synthesis of its parts" (A426/B454–55). In order for a whole to be given in intuition, it must be enclosed within limits which, even though they are not determined, are in principle determinable by the progress of intuition. A determination of this kind is not an instance of construction, since the limits themselves (which is what reason is interested in apprehending) are already fixed. If, however, the quantity itself is intuited, not the schema corresponding to a quantity, nothing except arbitrariness could limit such production. This should not be confused with the arbitrariness of the mathematical synthesis, in Kant's precise sense of the term. Rather, it means that the cosmological idea (in its possible accord with pure intuition) is generated in and by the motion of the regressive synthesis, not its possible coincidence or lack of coincidence with something given. If the mathematization of the cosmological idea is impossible for Kant, it is precisely because reason undertakes to construct a quantity of is own invention within the arena (pure intuition) wherein the construction of concepts is possible.

If this relation of the cosmological idea to mathematical conceptions is taken into account, then we can immediately brush aside the objection according to which Kant does not accept the possibility of *conceiving* of a process which would take an infinite time to *perform*. Kant's argument, so this objection states, simply rests on successive synthesis as the kind of extrinsic synthesis which we perform when we survey a spatial object part by part.[25] But this misrepresents the inner sense of the regressive synthesis. It is "motion, as an act of the subject . . . [which] first produces the concept of succession" (B154–55). If we were to embrace an infinite time in thought alone, and disregard the act of subject which sets the synthesis in motion, then we would not even be able to account for succession within that time.

The quantity of pure intuition cannot be measured since, from the point of view of the understanding in its empirical employment, it is equal to zero. Let us call the quantity of cosmological synthesis the *absolute* quantity, as opposed to the *arbitrary* quantity of mathematical synthesis.[26] It gives a transcendental representation of the infinite. Indeed, any representation of how great an infinite whole may be is impossible (A430–32/B458–60). Consequently, the infinite cannot be the object of a comparison between magnitudes. The infinite magnitude is that which is "greater than any number"—"greater" gives sense to the absolute quantity which cannot be measured at all. Thus, if the thesis could be demonstrated in an ostensive, non-apagogic manner, then the intuition of an indeterminate quantum as a whole should also fix the limits of the quantum. This is impossible, however, because pure intuition contains an infinity of partial representations. As the regressive synthesis is carried on, pure intuition takes in an ever larger number of conditioned appearances, so that the absolute quantity of pure intuition increases. But, on the other hand, as mere condition of possibility, pure intuition is always already infinite. In the constitution of new appearances as appearances of appearances, the transcendental condition of objectivity (namely, the reflection of transcendental apperception in the transcendental object) tends to shrink in favor of a more resolutely object-grounded condition. But if the balance tilts in favor of the object, is not the equilibrium between the subjective and objective components of knowledge so irremediably lost that the understanding might avoid illusion? No, because vesting pure intuition = 0 with absolute quantity is equivalent to bringing the transcendental conditions down to a minimum, that is, putting the understanding off its guard.

It was quite early in his intellectual journey, already in a pre-critical work of 1763, that Kant broached the question of the relation of the cosmological problem to the possibility of knowledge. Following mathematical methods concerning negative magnitudes, Kant opposed two kinds of nothing: the *nihil negativum* or logical nothing, which is absolutely impossible because it is contradictory, versus the *nihil privativum* or physical nothing, which, as Kant undertook to show, is present in the world. Kant's aim was to overcome the narrowness of traditional logic, which shuts itself up in the principle of contradiction, from which physical reality is supposedly deducible.

For the pre-critical Kant, the rebuttal of formally logical thinking happens to be based on mathematical truth. A physical nothing, he argues, can be compared to a resultant of positive and negative magnitudes. Indeed, there must exist negative magnitudes which are "something truly positive in itself."[27] This can be seen in mechanics, electricity, and magnetism, where oppositions of forces or poles appear to be mathematically measurable. Newton had stated quite clearly that "as in algebra, where affirmative quantities vanish and cease there negative ones begin, so in mechanics, where attraction ceases there a repulsive virtue ought to succeed."[28] Kant goes on to extend this idea to the cosmological problem, an extension that Newton had persistently refused on the grounds that the existence of a repulsive force could not be admitted beyond the confines of the internal structure of material particles. If the repulsive force could be metamorphosed into a cosmic force, so Newton thought, it would begin where attraction ceases, so that the observed global equilibrium in and of the universe would be explained without God's intervention. Instead of claiming that the force of attraction stops at a certain point, Kant maintains that, at the scale of cosmic magnitude, it becomes negative attraction. The result is that, when they are all taken together (added or subtracted, as the case may be) all real principles of the universe give a result equal to zero, so that "taken in itself the totality of the world is nothing beyond the fact that it is something according to the will of another."[29] Thus, the world negates itself without contradicting itself, which means that logical contradiction does not suffice to annihilate the world. On the contrary, the world draws its being from the very annihilation of contrary forces; this being is something, not by virtue of a certain rule of the understanding, but a *will* that is radically unintelligible to our understanding's reliance on judgments.

The fundamental implication of this enquiry for critical philosophy can be formulated as follows. As soon as we maintain a distance between our intellectual capacity, which is subjected to logical rules, and the being of things, the being of the world as totality is nothing. The real opposition of contraries at the cosmic level thus stands witness to this postulated distance. Since a will must give being, being, from the point of view of our understanding, cannot but present itself as always already posited.[30] In the language of the critique: when the understanding attempts to master the cosmological idea, the being of the world vanishes into nothing, so that the only possible universal foundation of knowledge must rest within the knowing subject. As regards the being of the world, this being which connects in its unity all the worldly objects, the subject can only form the concept of an object in general = *x*. This is the transcendental object, the unknowable part of the knowable, inasmuch as it empties itself of its being at the very moment when thought tries to relate to it. Just as, in the pre-critical *Attempt,* the world is that which includes all positive and negative appearances without having itself a being, so the being of nature (defined transcendentally as conformity to law) is the transcendental object, an *x* which makes the appearances possible as such. But whereas the pre-critical Kant was referring to nothing as a mathematical resultant, the mature Kant ascribes an absolute quantity to *x* which is not nothing, but which nevertheless can never be reduced to something mathematically determinable. Instead of referring being to an outside, as something posited beyond the concept, the concept of object in general fills the world with its own ontological vacuum.

This emptying of the being of the world by the concept receives its highest expression when the concept is none other than the cosmological idea. Indeed, the emptying lies at the heart of the process whereby the idea exhibits its allegedly synthetic character. Consider the highest principle of all analytic judgments. Kant claims that to formulate this principle as stating that "it is impossible that something should *at one and the same time* both be and not be" is to miss the mark. When the proposition is supplemented by a time condition, it asserts: "a thing = A, which is something = B, cannot at the same time be not-B, but may very well in succession be both B and not-B" (A152/B191–92). The time condition deprives the principle of its apodicticity. The predicate of the concept is no longer linked analytically to the concept of the subject, but synthetically (without our becoming aware of it), since the contradiction does not bear upon the subject but only upon the two predicates, inasmuch as they are posited at the same time. Now, the propositions of both the thesis and the antithesis in the cosmological antinomies must be synthetic in the following sense: Each demonstration rests essentially on the refutation of the opposite assertion, each proposition shows its object by disclosing a self-contradiction in the other object. Each proposition of an antinomy goes out beyond its own concept, only to refute the "appearances" referred to in the other proposition. But then, the apparently noncontradictory character of the remaining object does not suffice to ensure either its existence or its possibility. Kant had already argued in *The One Possible Basis* that the possible is not just the noncontradictory, it is also existence in noncontradiction—an existence that the possible presupposes. Transposed to the antinomy, this means that demonstrating that the other object contradicts itself amounts to demonstrating only that it contradicts *at the same time* the object to be shown. The predicates in each proposition are opposed to each another, if they are given at the same time, but this opposition could be eschewed if they were to be given *at different times*. In this case, they would constitute the concept of world by turning each of the cosmological propositions into an analytic judgment. The combination of the two judgments would make them appear to result from an achievable synthesis. The fact that the world possesses such or such a magnitude at any given time would then constitute the being of the world. Any two predicates of magnitude, even if one of them is infinity, would be compatible with one another if taken at different times.

Within the sphere of truth, the possibility of synthesis is already articulated in terms of a fundamental analytic principle. The synthetic unity of apperception is the highest principle of transcendental philosophy, in the sense that it makes possible, for all representations given in intuition, to be mine. But even though the principle unites all representations as mine, as principle it is analytic. Indeed, the "principle of the necessary unity of apperception is itself . . . analytic; nevertheless it reveals [*erklärt*] the necessity of a synthesis of the manifold given in intuition, without which the thoroughgoing identity of self-consciousness cannot be thought" (B135). Nothing (in particular, the connection between representations) is meaningful outside the mind, which therefore has a relation to any being because every being is within it—this defines its analytic identity. But every being does not yet make the totality of being. That is why the analytic identity of the mind draws aside in order to let synthesis accomplish its task, which is deployment of knowledge in time. Time means separation, incompleteness, and yet it is only in time that knowledge as a relation to the whole becomes possible. How does this relation of the analytic to the synthetic, which is the character of all transcendental truth, affect Kant's attempt to formulate a critical solution to the antinomies?

The Critical Solution, or the Doubled Illusion

In the pre-critical *Attempt*, the nothing as *nihil privativum* was the being of the world inasmuch as it is the resultant of the all the forces in the world itself. What is for pure reason the world without the being of the world? At first sight this would seem to be no more than a reaffirmation of the primacy of pure reflection: if totality is determined, then it vanishes as a totality; totality can be no more than an idea without determinate correlate in a possible experience. Yet, in the critical solution of the antinomies, Kant announces that the empirical concept becomes, once again, the standard of measure (*Richmass*). How can we understand this unexpected reversal?

The reversal expresses the need for a balance between a negative and a positive limitation of sensibility, without which knowledge would know of no limitation at all. Of the concept of a noumenon, Kant says that it is "a merely *limiting concept*, the function of which is to curb the pretensions of sensibility, and it is therefore only of negative employment" (A255/B310–11). This negative employment of the noumenon gives rise to a negative extension of the understanding itself, inasmuch as the "understanding is not limited through sensibility; on the contrary, it itself limits sensibility by applying the term noumena to things in themselves (things not regarded as appearances)" (A256/B312). How can the application of the understanding to things that are not appearances have a limiting effect on things that are wholly different, namely, the appearances themselves? By symmetry, when the appearances limit the infinite field of the understanding, they must give it its positive extension. How can the balance, let alone the encounter, between the positive and the negative be achieved? Only the idea of world is able to provide us with the necessary conditions. Indeed, the absolute totality requires an experience which will *always* be, in part, beyond the sphere of the sensible world. Therefore, as regards the totality, what cannot be an object of sensible intuition (noumenon in the *negative* sense) is related to the *positive* sense of the concept of noumenon, which is the object of a nonsensible intuition. The concept of noumenon in the negative sense occupies the place of the being of the world, the being that would be *if there had been no appearance at all*, that is, this being that is nothing as soon as there is sensible experience interfering with it. It results from this that while thought, in its relation to the transcendental object = x, empties the world of its being, this very same thought, in its relation to the unconditioned, is also a relation to the world *as if* the world were still filled with being.

Schopenhauer thought that he could make light of the cosmological antinomies, when he pointed out that there are no more than "a mere sham fight."[31] But Kant would have good reasons not to repudiate him entirely, since he himself asserts that each advocate, whether of the thesis or of the antithesis, "says more than it knows" (A472/B500). Indeed, if reason were to rely on the concepts of the understanding alone, disregarding the conditions of experience, it would obtain no more than "the general concept of a world in general," and in that case "no synthetic proposition, either affirmative or negative, can possibly be asserted" (A433/B461). We would be fated to remain at the middle point between affirmation and negation, which is also the zero point where the understanding contemplates its own infinite realm without ever going out beyond itself, were it not for the "passionate zeal" of the advocate of the thesis against the "calm assurance" (A465/B493) of the advocate of the antithesis. In the antinomy, one blows hot and cold for the sole purpose of disturbing the infinite rest of a satiated understanding, which invites reason to dogmatic torpor. The balance between negativity and positivity (which represents the givenness of being) must be shaken up, so much so that critical reason finds itself caged in a chasm. Without this quickening, thought would always run the risk of settling peacefully into being before the infinite race for the possession of the unconditioned can even be run.

This requires that we draw up not only an opposition between statements concerning being, but also a correlative opposition between modes of access to it. Thus, in the antithesis of the first antinomy, Kant refuses to include in its proof-structure any demonstration of the infinity of the world that would reject the finite world on the basis of some putatively impossible intuition. *Intuition* is the procedure of the thesis, whereas the antithesis can rely only on the *concept* of an impossible relation. As a result, the possibility of a direct or ostensive demonstration of either proposition must be rejected. In the observation on the antithesis of the first antinomy, Kant develops his example of an attempt to eschew the apagogic mode of proof of the thesis: the argument is "that a limit of the world in time and space" could avoid the obnoxious consequence "of having to assume that . . . the infinite void must determine the magnitude in which actual things are to exist" (A431–33/B459–61). Kant rejects this argument on the grounds that it proceeds from a doubly surreptitious substitution: for the first beginning (an existence preceded by a time of nonexistence) an existence in general is substituted, which presupposes no other condition whatsoever; and for the boundaries (*Grenzen*) of extension limits (*Schranken*) are substituted. By limit, Kant understands that which encloses, circumscribes, and which is itself limited by something else: "in all limits there is something positive (e.g., a surface is the limit of corporeal space, and is therefore itself a space)." By contrast, boundaries "contain mere negations" because what lies beyond them is nonbeing.[32] And yet, in order to show that this conception is unacceptable, Kant proposes what seems to be an even more paradoxical relation: "it cannot be denied that these two non-entities (*Undinge*), empty space outside the world and empty time prior to it, have to be assumed if we are to assume a limit to the world in space and in time." If the limit of extension is itself limited by a nonentity, it seems that this limit has a great deal in common with a boundary. Is it not paradoxical, then, to say of empty space and empty time that they are nonentities? The conception to be rejected is less excessive, since it converts each nonentity into nothing, that is, a negativity into a zero.

This latter conception is not satisfactory, for otherwise an ostensive proof of the thesis would become available. Thus, curiously enough, in order to refute any direct demonstration of the thesis, one must rely on what the antithesis asserts. One must, in fact, go as far as exhibit this assertion as if it were true. The real issue is to determine what is "true" in the antithesis, without, in so doing, falling into a direct demonstration of the antithesis. Once more, intuition must be brought in, the very intuition which is excluded from the proof of the antithesis as it is explicitly given. Instead of an outright conflict, we should certainly be right to speak at this point of a show of conflict. On the one hand, a direct proof of the antithesis based on intuition is impossible by virtue of the whole proof-structure of the antinomy. On the other hand, such a proof would seem to be required by the need to refute any possible direct proof of the thesis. In the direct proof of the thesis, the substitution of boundary for limit does not even come to grips with the time and space of a sensible world. The world with a boundary is a purely intellectual world, which therefore proves *more* than what was originally intended in the thesis: it would prove that the finitude of the world is also a guarantee of the finitude of experience within the world. On the other hand, if there were anything like a proof of the antithesis in accordance with intuition, this proof would begin by applying an intuition to the intellectual concept of finite world. It would then show the impossibility of such an intuition, which is just what the direct proof of the thesis requires. Ultimately, what differentiates the direct proof of the thesis from the proof of the antithesis in accordance with intuition is this: if the groundlessness of the former is established, we are *not* sent back to the latter, since the intellectual concept of a finite world does not lend itself to the apagogic mode of proof—it is not essentially a cosmic concept.

As the transcendental illusion deploys itself in the apagogic mode of proof, it thus remains safe from all assaults. If Kant's argument succeeds, then there is no "truth" of any sort that is able to counteract a transcendental illusion, precisely inasmuch as such an illusion passes off for true. Illusion can be disclosed only by applying to it a new negativity (a new illusion), an illusion which plays into the hands of the transcendental appearance in order to exhibit it as such, without being hampered by a noncosmic concept. Following what he takes to be the empirical foundation of Epicurus' teachings, Kant approaches this new kind of necessary illusion in the following terms: "That, in explaining the appearances, we must proceed *as if* the field of our enquiry were not circumscribed by any limit or beginning of the world" (A471/B499, emphasis added). This is the precondition for the critical solution. The antithesis must be repeated, so that it can appear as a deliberate illusion—a product of reflection guided by no determination. That is, the antithesis must be considered as if it were true, which turns it into the illusion of an illusion. In this sense, Schopenhauer's description of the antinomy as sham fight really applies to the critical solution, not the antinomy itself. Thinking in the mode of "as if" is a new experiment with concepts, where the critique experiments with itself, instead of demarcating itself from the precritical or non-critical modes of thinking. Because the entire critique is involved in this experiment, not just its starting point, the already established truths within the realm of possible experience are themselves under threat. Likewise, the entire critical solution must be read in terms of the doubled illusion. Kant's text becomes extraordinarily intricate, abreast of the difficulty of the thing itself.

The first effect of this new experiment is to drive us back to the critical correlation between analysis and synthesis. Kant asserts that "since we are here dealing solely with a thing as object of a possible experience, not as a thing in itself, the answer to the transcendent cosmological question cannot lie anywhere save in the idea" (A479/B507). Because it is related to a possible experience, cosmological reasoning would seem to be synthetic. Yet the object is not really outside the concept, since the idea itself "is a mere creature of reason," which gives its reasonings the look of analytic judgments. Regarding the absolute whole of quantity (the universe), however, "all questions as to whether it is brought about through finite synthesis or through a synthesis requiring infinite extension, have nothing to do with any possible experience" (A483/B511). But is not the distinction between a finite and an infinite synthesis crucial with respect to the temporal essence of any synthesis? The critical solution puts us on the way to a possible experience that is no longer determined absolutely by time. As we try to explore the antithesis as though it were true, we find ourselves storm-tossed, from one extreme to the other—from the synthetic to the analytic and back. Contrary to analytic identity, not every being is within the mind, since the regressive synthesis, at each of its moments, has *not yet* run through every being. In accordance with this identity, however, the idea relates to the totality in such a way that every possible being is included in it *in advance* of actual experience. When Kant writes that "the solution of these problems can never be found in experience" (A484/B512), he does not mean that we should forget about the constitution of the cosmological idea through appearances of appearances, but rather that we no longer know for certain whether the synthesis is identical to itself as we move in it. The regression might have to be pushed so far away from its starting point that the starting point becomes unrecognizable. Nature could be so overwhelming that we will make no progress by supposing "the whole of nature to be spread out before us, and that of all that is presented to our intuition nothing is concealed from our senses and consciousness" (A482/B510); nothing excludes that the supposition might be crushed by the sheer weight of the whole of being.

Even though its relation to the object is more intimate than in the case of the category, the idea of reason does not make "truth" possible because it does not account, from within itself, for its relation to the object. The question is one of outlining the origin of the transcendental illusion, knowing full well that this origin is concealed by transcendental illusion itself (since the idea of the world is never completely identifiable with a pure concept of the understanding). The only possible way to go about this consists in producing this origin as a *new simulacrum* (which Kant refers to in terms of "as if") without which the effects of previously constituted simulacra could never be annulled. This is the price we must pay in order that reason might allow itself to change its final end, that is, to substitute the totality for the unconditioned; we now see that this change is itself the effect of the new experiment by means of the redoubled illusion. Initially, the change of ends was itself imposed by the need to curtail the operations of the imagination beyond the realm of possible experience. Thus we should not be too surprised at the brutal reversal Kant imposes on the conclusion that has just been formulated. He announces that "to obtain absolute totality in the *empirical* synthesis it is always necessary that the unconditioned be an empirical concept" (A487/B515). Kant had said, prior to bringing the antinomies into play, that the idea of world would in no way be affected by the effects of its problematic relationships to empirical concepts; the latter are never perfectly adequate to it. But as the critical foundation of the antinomy is about to be disclosed, the empirical concept becomes, once again, the standard of measure (*Richtmass*). This is not to mean that we are simply going back to what the understanding in its empirical employment teaches us, for otherwise the antinomy would have been useless for the critique. Since it always deals with a possible object of experience, the idea is necessarily always too small or too large for any concept of understanding, according to whether the world has a beginning (a limit) or not. It is thus not only empty *but also* devoid of meaning (A486/B514). The concept without intuition is empty, whereas the intuition without concept is meaningless. As a result, the idea gathers together, simultaneously, all prerogatives of the empty concept and the meaningless intuition.

To be sure, when the idea is too large for the understanding, it conforms to reason. But all the same, it is the understanding that pushes the finite part of the world that has already been surveyed to always enlarge itself. This is ultimately the reason why the transcendental foundation of the antinomy must lie in what is "true" in the antithesis. Instead of saying that the empirical concept is always too small for the idea of a world which has no beginning, Kant constructs his new illusion carefully enough so as to make it appear that such a world is too large for our concept (A489/B517). This enables him to *simulate* an objective validity for the idea; *only later*, when he has shown that no empirical regress can be found to correspond to it, does this idea acquire a significance of its own. If, on the other hand, we had started from the regress, in order to show that the concept is too small for the idea, we would never even have reached the idea, that is, the transcendantal simulation of the whole cosmological idea would simply have been left in place. Having thus made the effect of the newly fabricated illusion stand in relief, Kant is now in a position to conclude that it is "at least a well-grounded suspicion that the cosmological ideas, and with them all the mutually conflicting pseudo-rational assertions, may perhaps rest on an empty and merely fictitious concept of the manner in which the object of these ideas is given to us" (A490/B518). The antinomies were initially presented as rational, but the new illusion casts doubt on this rationality, thus transforming the natural dialectic into a mere sophism which can be done away with. Indeed, the empty point of contention involved in the new illusion is not the object of the idea, but the mode of givenness of the object; this places us at the heart of dialectic as a play of "given" illusions (analogous to optical illusions) and "contrived" illusions of illusions. Thus, in one of Kant's later expositions of the same argument, the emphasis on the constraint involved in the illusion of illusion becomes so clear that he no longer hesitates to qualify the proofs of the Analytic as "dogmatic," thereby reversing the roles usually ascribed to the Analytic and the Dialectic. As he puts it: "The antinomy of pure reason unfailingly leads back to that restriction of our cognition. Here in the Dialectic what was proven dogmatically in the Analytic is likewise incontrovertibly confirmed by an experiment of reason that it employs upon its own capacities."[33] Kant now presents his experimentation with concepts as lying entirely within the illusion of reason; the contrivance whereby an original illusion is turned into the illusion of an illusion is the core of this experimentation.

The effect of this force on the original antinomy is to solve the interior tension from within, freeing this tension outward onto a new object, the correlate of the new illusion. Indeed, at the limit-point to which any regressive synthesis can be pushed, the transcendental object (which orders the concatenation of all our possible perceptions) loses its natural connection to experience and finally becomes indistinguishable from an in-itself, since at that point it is possible to refer to it as being "given in itself prior to all experience" (A494/B522-23). Had such a confusion between the transcendental object and the noumenon been with us since the beginning, the whole Copernican revolution could not even have started; but at this point, the confusion has a liberating effect. It permits us to remove the ambiguity (which has been with us since the statement of the proof of the thesis) which holds sway over the impossibility of having the experience of an interval of past time in a present time. We can now convert this impossibility into the index of a possibility, the condition of which lies in my own present existence. All that is needed is to accept that the desire to retrieve the unique direction of time from past to future can be definitively abandoned. Consider all those events which have taken place prior to my own existence. Their significance for the present is that they are "nothing but the possibility of extending the chain of experience from the present perception back to the conditions which determine this perception in respect of time" (A495/B523). Kant argues here for the first time that a coming to presence of what lies entirely in the past, relative to a present, is an extension or amplification from that present, not an unfolding from the origin of the world. Significantly enough, it is the passage to the in-itself which supports the possibility of such a synthesis.

We are now at the point where we can enter the critical solution to the cosmological problem. This solution has to be viewed as the ultimate expression of the problem of synthesis. Pure reason constructs its dialectical argument in the following way:

> If the conditioned is given, the entire series of all its conditions is likewise given; objects of the senses are given as conditioned; therefore, etc. (A497/B525)

Kant endeavors to show that between the major and the minor premise of this reasoning there is inconsistency with respect to the object. For the following paradox is immediately apparent. Pure reason creates its own object "analytically," but in order to remain embedded within the realm of appearances, the only possible concept capable of generating such an object must be a pure concept of the understanding. How can an analytic proposition account for a knowledge which has as its foundation the very possibility of an a priori synthetic knowledge? This becomes possible, Kant argues, if the major premise is replaced by the following proposition:

> if the conditioned is given, a regress in the series of all its conditions is *set* us *as a task* [aufgegeben]. (A497–98/B526)

This amounts to bestowing a heuristic value upon the analytic proposition, or, equivalently: the cosmological object is an object of synthetic knowledge, but only a posteriori. Viewed from this perspective, the cosmological problem is finally reduced to an a posteriori reflection concerning an a priori that escapes us radically.

In the unrevised major premise of the cosmological reasoning, the connection between the given conditioned and its given condition takes place in such a way that they present themselves without any order of time—if they are given, they are given at the same time. An antinomy appears, then, because the regressive synthesis is undoing what has always and already been done outside the concept, and this undoing takes time. In its unrevised form, the major premise should thus constitute a true analytic proposition, inasmuch as such a proposition is simply formal, the time condition being superfluous. But this cannot be the case, as we already know: this proposition is *not* analytic in the sense of being clearly distinct from the synthetic, because it must be used to progress toward antecedent conditions which ultimately exceed the realm of appearances, even though it never goes out beyond itself. Therefore, it is not strictly speaking accountable for the logic of contradiction. There must be another logic specific to the cosmological reasoning, which Kant articulates in such a way that two judgments which are opposed to one another need not be contradictory.

The two opposing propositions are quarrelling over nothing, in the sense that each denies the other without affirming anything determinate. The new simulacrum, however, forces them to affirm what they do not affirm in the original antinomy, namely, that the universe *is* empirically either infinite or finite. Now this magnitude which is affirmed must be *inessential* to the concept of universe, for otherwise we could not understand how what is denied by one proposition in the opposed judgment remains attached to it; it remains attached even though the whole concept is denied, precisely because it is inessential. In other words, both propositions have failed to take into account a third possibility, namely, that the concept of world might be perfectly consistent even though it has no magnitude at all. As a general case, Kant considers the pair of judgments "all bodies are good-smelling" versus "all bodies are not good-smelling": both might be false since it is possible to say of a body that it has no smell at all (A503/B531). Magnitude turns out to be an *accidental condition* of the concept of world, whereby a deceivingly synthetic character is attached to it. In the antinomy, each proposition contradicts the other essentially because, in the apagogic mode of proof, only an accidental condition of the concept is suppressed. Likewise in the critical solution, if a regressive synthesis toward antecedent conditions is merely proposed or set as a task, this means that such a synthesis is merely offered to us: we remain free to take it up or to turn it down.

This explains how Kant proceeds to his decisive move, namely, the connection between magnitude of the world and determination in itself. Suppose the antithesis is true. In this case,

> if we had said that the world is either infinite or finite (non-infinite), both statements might be false. For in that case we should be regarding the world in itself as [*als*] determined in its magnitude, and in the opposed judgment we do not merely remove the infinitude, and with it perhaps the entire separate [*abgesonderte*] existence of the world, but attach a determination to the world, regarded as a thing actually existing [*wirklich*] in itself. This assertion may, however, be likewise false; the world may not be given as a thing in itself, nor as being in its magnitude either infinite or finite. I beg permission to entitle this kind of opposition *dialectical*. (A504/B532)

Retrospectively, the proofs given in the antinomy can be reinterpreted as not being baseless illusions, because they were grounded "on the supposition that appearances, and the sensible world which comprehends them all, are things in themselves" (A507/B535). Kant distinguishes between two cases: the analytical opposition, in which the falsity of one proposition entails the truth of the other, and the dialectical opposition, in which both propositions are false. The new experiment with concepts—taking the antithesis as if it were true—enables us to find out the false presupposition underlying both propositions in the dialectical opposition: that the absolute totality of things is a thing in itself. If the presupposition is accepted, as in the antinomy itself, then the opposition becomes analytical. The possibility remains not to accept the presupposition, in which case the regressive synthesis is merely proposed as a task. But where does the presupposition itself originate? Since they were originally based on an amplification of the (dynamical) law of connection between perceptions, and since any such necessary connection is a determination of existence, both the thesis and the antithesis seek to admit the *existence* of a world, either finite or infinite. But, as Kant explains in the section on the impossibility of the ontological proof of the existence of God, existence is not a real predicate, in the sense that "it is not a concept of something which could be added to the concept of a thing" (A598/B626). To add a determination of existence is simply to posit something with all its predicates, irrespective of whatever and however many predicates we think in this thing. In reference to the first antinomy, Kant speaks of *separate existence* when referring to the operation whereby a certain magnitude is first removed and then replaced by another. That is, when the transcendental illusion is constrained to lay bare its existence as a result of its coming up against the new illusion fabricated for the purpose, we find that what each proposition should have affirmed explicitly is that existence with a certain magnitude *is* a real predicate, that is, a determination that is added to the concept of world.

The fundamental implication of the critical solution is that, once the world is deprived of magnitude, it also loses its character as an unconditioned whole (A505/B533). If the world is something unconditioned, no magnitude can correspond to it because it is in itself. But then, it could not be a totality either, which contradicts the idea of world. Thus either the idea of world is meaningless, or any series of appearances must always be conditioned, which contradicts the notion that the world contains all that exists. The very idea of world tends to vanish as a result of the dialectical opposition. It is not true that the world is a whole existing in itself; but if it does not exist in itself, it is also not a whole. Now, even though the regressive synthesis can never contain itself in its entirety, suppose that the world must, in one way or another, possess a magnitude inasmuch as it continues to be a whole. Is not the claim that the world has no magnitude tantamount to saying that the world has a magnitude, but that the synthesis corresponding to it can only be posited? Kant's cosmological principle of totality (*kosmologischen Grundsatz der Totalität*) (A508/B536), which was already presented in chapter 3, comes at this point as a final expression of the supposition that a whole exists. It prompts the regressive synthesis to continue endlessly beyond possible experience, precisely because it is torn between the world within (the sensible world of experience) and the world without (the world in itself). In this way, the common presupposition underlying all dialectical opposition (that the totality is a thing in itself) is and is not accepted at the same time. If empiricality is to become the standard of measure, however, the world without must somehow be made comparable to the world within. This is conceivable if, and only if, the portion of the world accessible to experience at any given time is supposed to be representative of the world in its totality. There is still the problem, then, that what it means for a portion of the world to be representative of the whole may change with the progress of intuition. Allowing for such a change implies, in turn, any absolutely fixed unconditioned to be completely banished from *both* the world in itself and the experience we have of it. Thus, even if the totality is given in an empirical intuition (second antinomy), the regress continues up until infinity, without ever encountering an unconditioned. Likewise in the first antinomy, where no such empirical intuition is available, the regress can be regarded as indefinite (A512–15/B540–43). In fact, in the latter case, the idea of totality *plays the same role* as the empirically given totality: both are equally impossible.

Following Kant's argument in the introductory sections of the Dialectic, we began by realizing that the cosmological idea is nothing other than the putatively objective correlate of the transcendental unity underlying the totality of all possible appearances. The idea is supposed to provide the sensible figure that the transcendental object would assume if such an absolute knowledge were actual. Because it is overwhelmed by the weight of being, the integral intuition of everything would still not be enough to guarantee the coincidence of the transcendental object with the absolute totality. So that "even if the whole of nature were revealed to us, we should still never be able to answer those transcendental questions which go beyond nature. The reason for this is that it is not given to us to observe our own mind with any other intuition than that of inner sense" (A278/B334). In the absence of any unconditioned beyond the limits of experience, we are sent back to the origin of our own sensibility. The form of sensibility is that in and by which any vision of things is given. Certainly, it also has a vision of itself by means of inner sense; but time itself cannot be perceived, that is, this vision of itself is in principle subject to, and controlled by, the understanding, which places it entirely in the service of a determinate knowledge of the manifold of intuition. However, when the influence of sensibility on the understanding is unobserved, the determination of a manifold beyond any possible experience of nature is also a necessary illusion, in the sense that that in which all vision takes place becomes visible—the "absolute quantity" of pure intuition. The transcendentally ideal character of inner sense emerges strenghtened by the trial which it has just undergone by involving itself in the colossal sham of its provisional abandonment, strenghtened to the point that it remains as the only possible vision.

Dialectic without Nihilism

The antinomy of pure reason can oppose intuition and thought so saliently—as though the one were completely dissevered from its relation to the other—precisely because the Idea of reason brings them as close as possible one to the other. That is why, in the final analysis, the Idea can be said to be a problem, both before and after the exposition of the antinomy. Does that mean that, in reality, we never leave the problem behind, that we have just revolved in a circle without getting to the bottom of the difficulty? Does admitting the antinomy as a conflict of reason with itself force us simply to retrench ourselves in the problem of reason contemplating itself, so that the antinomical conflict is just the expression of this problem? It is one thing to show that reason gets caught up in a problem involving itself, but it is quite another thing to show the deep root whence this problem springs.

According to Heidegger, the ground of Kant's system of reason remains ultimately obscure. Kant has not shown the origin of the ideas, that is, the ground of the system.[34] A hiatus crops up between the total system of reason (the ideas) and beings as a whole because, as he puts it, the ideas "do not present what is meant in them." Their function being regulative rather than constitutive, they "are only directions for finding, but themselves are not found." Heidegger's interpretation is based on his own hermeneutic approach to being: he refers to the critical solution of the antinomies in order to make sense of the original ideas that give rise to antinomies. There are two problems with this interpretation in the present context. (1) It is implicitly assumed that a smooth transition connects the presentation of the antinomies to their critical solution, as if, retrospectively, one global dialectical structure could represent the entire work of reason.We have seen that the transcendental logic of illusion is quite the opposite: the transition to the critical solution requires that we follow the bent of Kant's own personal genius, which takes us away from any fixed structure of dialectical reason. (2) If the ideas do not generate any concept, this is precisely because they are closest to the originary relationship of thought to intuition. *The specific difficulty of Kant's Dialectic—but also its unique character—is that the ideas give us to see the generation of meaning without any new concept being thereby generated.* When the withdrawal of thought with respect to the world (categorial constitution of experience) is complete, the withdrawal is converted into a transgression.

We reach here the point of highest contention regarding the destiny of the entire critical work. When Heidegger argues that the whole of reason is characterized in terms of one global dialectical structure, and when he goes on to discover that this structure cannot stay in place because it gives only a direction for finding (the whole of being does not ground the system), he echoes the tendency of associating critical philosophy with what has become known as "nihilism." The tendency has become quite common, ever since Jacobi expressed his reservations concerning Kant's critical philosophy in a famous letter to Fichte of 1799: when the self posits itself as identical to itself by excluding the primary relationship to the whole (Copernican revolution), and when later, in the Dialectic, the self can then only attempt desperately to reconnect itself to the whole, this attempt is, of course, doomed to fail. The dice are loaded: the rupture with being is, so to speak, consummated in advance. In his interpretation of Kant, Heidegger expresses this negativity by saying that, as Kant goes from the Analytic to the Dialectic, "Kant himself undermines the floor upon which he initially placed the Critique."[35] Indeed, the concept of pure reason and the unity of a pure, sensible reason become *problems*, they lead us to *darkness*. This conclusion is true only as long as the Dialectic follows literally the Analytic. If, from a phenomenological perspective, the Dialectic opens up the whole critical project, then the darkness of the idea of reason is merely a consequence of the highest proximity that it brings about between thought and intuition. Heidegger describes Kant as "falling back before the ground which he himself unveiled," and interprets this falling back as "that movement of philosophizing which makes manifest the breaking-open of the foundation." Heidegger does not see that the paradox of the *Critique* is that the ruin of the foundation follows only in the act of encircling this foundation. The Dialectic is not the undermining of a previously established foundation; it only enables us to reflect more directly than anywhere else upon the problematical foundation which guides the critical enterprise from the beginning.

Therefore, all the nihilistic consequences usually ascribed to the *Critique* will follow if and only if it can be maintained that thought, conceived as pure reason, is, in effect, a thought given over to its own internal systematicity. Now, Kant has just called "dialectical" the opposition of two propositions that returns us *below* the antinomical opposition properly so-called, as if the dialectic were to begin only at this point. It all looks as though the provisional abandonment of the transcendental ideality of space and time had not as yet placed us within the dialectic, which begins only subsequent to the effort toward its re-establishment. The freedom of the critical philosopher—Kant's personal stroke of genius that makes us see the illusion with the eyes of inner sense—is not a negativity, or at least not yet. The negativity of pure reason remains to be understood more properly: as it turns out, it is found outside pure reason altogether, as if the freedom of the critical philosopher had been no more than a preparatory step in this direction.

CHAPTER 6

A Reversal of the Reversal

Crossing the Border of Reason

Existence is not a real predicate. The cosmological significance of this discovery is that the copula "is" in any judgment of the form "subject is predicate" is the true center of the judgment. But whereas the vision of being can see itself, and thus center itself in its own formal infinity, being cannot see itself as center; it becomes unreal when it does so. Hence, even if there were a privileged standpoint from which the totality of existence could be judged, this point could never be determined. Differently put: the Copernican revolution is limited to the model of the solar system. However, the question remains: Is there not a positive interpretation of this phenomenon? The center is displaced as the system is enlarged. If the system were to be the absolute totality, then the center would be éverywhere. In an effort to align the center of vision with the periphery of being, could we not say that existence is simply spread over the whole of the judgment? If this were possible, could we not once again speak about the world as an unconditioned whole, even though its actual magnitude can never be determined? This last effort forms the backdrop of the post-Kantian attempts to complete Kant's system.

Kant's examination of the cosmological problem has led to the rejection of both mathematics (as a mode of access to the unconditioned) and the unconditioned itself (as the final object of reflection). Thinking the absolute totality is thus the critical test of the whole *Critique*—a counter-experiment with concepts—because it dangerously combines the two movements of thought that critical philosophy is eager to distinguish from one another: the movements "downstream" and "upstream" from a priori knowledge, that is, mathematics and the transcendental ideal respectively. The early history of post-Kantian philosophy reveals how much Kant's attempt to combine these two movements of thought has not been genuinely pursued. His immediate followers departed from him and bequeathed to us two totally divergent directions. It is crucial to track down these two directions, best represented in the works of Maimon and Fichte, not only in order to further elaborate the essence of Kant's project, but also in the hope that a realization of its fragmentation will give us an indication of what has become of what Kant called the "shining misery" at the dawn of modern times.

Perhaps these two movements of thought have not been combined because they could not be. We face a phenomenon peculiar to the Kantian dialectic, which leaves the cosmological principle in a very precarious position. Since reason needs to transgress the limits of possible experience the idea of absolute totality disintegrates. Since there is no difference between an idea and a category when neither has a corresponding object, the transgression would seem to lead to the collapse of the fixed structure of the categories. Hegel would see this as the point at which the logic of illusion forces upon us a division of the rational concept itself into two branches, each branch being a *new* concept with its own conclusions. Such an interpretation requires that the antinomy should reside in the concept itself, not in the relation of the concept to intuition.[1] Despite what Hegel says, the Kantian sense of contradiction of reason with itself is not fully internal to reason. Rather, it lies in the relationship of reason to its object in the world. This relationship becomes apparent only when the object *is* the world, so that the solution to the contradiction implies a discourse *external* to the dialectic. Since there is no privileged standpoint from which this external discourse is possible, it must use the figure of the doubled illusion. This distinction vanishes in Hegel's dialectic, which starts from an essential contradiction of reason only then to move on to the idea of an essential contradiction in the object itself. The solution can thus become internal to the dialectic. For Hegel, two magnitudes for the totality lead to two ideas of reason, that is, when the universe has been thought twice, the whole of thought has not yet been exhausted. For Kant, however, the universe as object absorbs the whole of thought (it is an object of pure reflection, simultaneously the same and different), inasmuch as it cannot be thought along with anything else.

Kant's cosmological principle, which Hegel apparently ignores, is thus the strategy adopted by reason in order to escape the baleful collapse of the system of categories, and thereby to surmount the contradiction of a non-mathematizable quantity. But Hegel's interpretation itself results from the recognition of a division within a priori knowledge, which was prepared by Kant's immediate successors.

Ironically, whereas Kant sought to complete his critical philosophy by examining the limits of the object *in* the idea of absolute totality, it was precisely the attempt to complete the idea underlying the whole Copernican revolution that led to its alteration. Heidegger would ultimately express the consequence of the recasting of Kant's Transcendental Dialectic in relation to the third cosmological antinomy—the conflict between freedom and necessity. Since Heidegger takes the ultimate horizon of critical philosophy to be the truth of Being, it is not the object of philosophy, but philosophy itself that becomes "intrinsically a strife between necessity and freedom," in the sense that "it will produce from itself this strife."[2] With Kant, philosophy is still *about* the conflict, but being comes into play in the mode of unobserved influence. This being is a transcendental appearance, that is, the world-horizon connected with the subjective conditions of knowledge. Now, if this mode itself gets thematized, then, as we have seen at the end of the last chapter, philosophy can only double the transcendental appearance—it does not insinuate itself into appearance to become the expression of the conflict.

For Kant, the old dogmatic dreams of reason are the result of a much more profound situation of thought, one which only critique really addresses, namely, the phenomenality that constitutes all thought in its original relationship to the world. Once thought seeks to investigate its own origin *in* the world, this phenomenality becomes unmasterable. The only way to combat these old dreams is to invent an illusion of the illusion, to duplicate the illusion in which the antinomy would have trapped us. This doubling is the final attempt of critical reason to master the phenomenality of thought. This attempt is made possible by the fact that the absolute totality of conditions (the universe) is a very peculiar object; it can be thought only in the mode of pure reflection, with no prior determination. We can get a sense for the difficulty of this undertaking by examining its fate in Schelling's philosophy. According to Schelling, the illusion of illusion cannot return us to the ultimate ground of thought. Indeed, he argues that *mere* reflection, that is, reflection disconnected from the determinations of the world, "is a mental disorder in man."[3] Just as Kant had argued in the *Anthropology* concerning the mathematical mind, Schelling goes on to explain that such a disorder takes control of the whole person who is subject to it. Because the resulting chimeras lie beyond all reason, no battle could ever be fought against them. Likewise, for Kant, there is no antinomy in mathematics. Following Schelling, mere reflection, which takes us into the illusion of illusion, brings us out of the realm of reason.

Nothing can be said about the a priori as such. The chief lesson of the whole of Kant's critical enterprise is that we can only try—indeed, we cannot help but try—to speak a posteriori about the a priori. Once one says something about the a priori as such, one makes it an in-itself. Thought must relate to an outside, to the world, without knowing what this outside is. This is the ultimate meaning of the Copernican revolution that Kant initiated; only by positing an outside can thought "find itself" in itself, but then it never really adheres to itself again. What allows knowledge to reflect upon itself and thus to legitimate itself is that, within knowledge, there is an origin which escapes it. For thought, there is no origin of thought, and yet there clearly must be one. As a result, the legitimacy of knowledge founded on thought cannot be attached to something other than knowledge itself. This does not mean that transcendental knowledge will eternally contemplate itself. What could legitimate progress, or the movement of knowledge out of itself? We are forced to put the cart before the horse, as it were. An illusion always lies *before* knowledge. This is the necessary illusion of attributing ontological status to knowledge. What is thus ontologized is put *in the place of* the unknowable origin. But could we not locate this origin in the "outside" which thought has expelled from itself? Attempting in this way to recapture the unity between itself and the world, thought engages in the antinomies of the Transcendental Dialectic, which form the very place in which this rupture is reflected in the world.

Let us first consider Solomon Maimon, whom Kant himself hailed as one of his most perceptive interpreters.[4] Maimon wants nothing less than to restore the status of mathematics by eliminating the condition of merely relative perfection that Kant had assigned to it in the *Critique of Judgment*. He is quick to forget Kant's warning that the transcendental ideal is the product of a natural idea, as opposed to the merely arbitrary idea that lies at the basis of mathematical synthesis. Indeed, Maimon is content to conflate the two: "All of the concepts of mathematics are thought by us and at the same time exhibited as real objects through construction a priori. We are, in this respect, similar to God."[5] His starting point is a complete renewal of Kant's notion of intensive magnitude. Kant determines empirical reality a priori with regard to *form* alone, except in the "Anticipations of Perception," where intensive magnitude is a quality defined as the intensity of sensation, or the degree of influence on sense (A166/B208). This degree can always be diminished in thought, but it can never vanish completely. Otherwise, reality would be indistinguishable from its own absence, that is, negation. Kant explains: "Between reality . . . and negation there is therefore a continuity of many possible intermediate sensations" (A168/B210). Consciousness is suspended between givenness and nothingness. Between them, there is realm of possibility that remains pure, in the sense that it never becomes an object of apprehension even though it is the condition of possibility for any given sensation to be apprehended. Thus, it is the *matter* of intuition that is foreclosed by means of the intensive magnitude (A166–68/B208–10). Maimon reads this argument in the following way. If what is given in sensation was prior to its apprehension in terms of a succession in time, and if this given left no trace in consciousness, then there must have been some kind of conspiracy going on behind the back of sensation.

Maimon ascribes the intractable action of matter to the limited (or unconscious) activity of consciousness. According to him, this activity reveals the presence of a residual quantity that no quality is able to absorb. Concerning the intermediate sensations that remain in the mode of possibility at all times, separating givenness from nothingness, Kant had said that "the difference between any two of [these] is always smaller than the difference between the given sensation and zero or complete negation." Because the "Anticipations of Perception" is one of those transcendental principles of pure understanding proceeding from a concept to an intuition, instead of from an intuition to a concept (as in mathematics), no mention was made of the mathematical concept of differential. But Maimon uses the mathematical concept of differential—that is, a quantity that is itself smaller than any assignable quantity—in order to account objectively for the progression, within sensation, from zero to its infinitely small degrees. Maimon needs to find a counterpart for this silent activity of the mind in the object itself. He finds that if our sensibility furnishes differentials for a determined consciousness, then the differentials of objects themselves cannot be encountered passively by our consciousness, but are produced by its transcendental imagination. Since the real is now a function of imagination, this solution would lead to an outright skepticism were it not for Maimon's following argument, according to which the unsensed differentials of objects account for nothing less than the things in themselves.[6] Indeed, objectivity would vanish in the absence of any reference to the noumenon, since the jolt provided by external reality would not be needed if consciousness had only the result of unconscious syntheses before itself. Consciousness would never be able to discover a necessary connection of its representations to itself were it not for mathematics, which provides us with the example of such a necessary connection in which the independent content of the appearance is annulled.

This is the intriguing result of Maimon's investigation: in order to be completed, the mathematization of the forms of sensibility requires that the noumenon be rehabilitated. The perfect infinity of these forms is now an Idea that has no significance for the way the faculty of concepts synthesizes a manifold of intuition, precisely because this manifold is completely transcended. Instead of Kant's two beginnings (the given sensation, on the one hand, and its matter below givenness, on the other hand), Maimon can argue that "our knowledge of things begins at the middle and also ends at the middle."[7] This is what occurs in mathematical knowledge. When we start from a certain magnitude we can proceed either backwards or forwards by following the same rules of unity, and so we can always think a unit as being larger or smaller than anything given in sensory experience. Arbitrariness has been transposed from the mathematical synthesis to the beginning of experience itself. The way is clear for overcoming Kant's claim that intuitions cannot be reduced to concepts. After Kant, as the history of mathematics has shown, all the logical problems that he thought to be irresolvable as long as this irreducibility was denied, have been resolved by means of the imaginary variable that Kant rejected as part of mathematical science itself.[8]

Undoubtedly, Kant's Transcendental Dialectic contains an implicit response to Maimon's development. The theory of the unobserved influence of sensibility would enable Kant to resist Maimon's suggestion that the matter of sensation is posited by the activity of the ego. Instead of being an unarticulated accompaniment of the activity of the mind (the barely perceptible signals we receive once we shift into a constructive mood and thus liken ourselves to God), this matter would be ascribed to an excess of appearance that happens not to be fully mastered by the mind. The experience of such a surplus forces human reason to seek the unconditioned *within* the series of conditions of the phenomenal world. But the post-Kantian attempts to silence the Dialectic did not come only from below (mathematics), but also from above (the transcendantal ideal).

Indeed, at the other extreme from Maimon we have Fichte, who returns to the ambiguous principle of reflection in transcendental philosophy, which states that a first reflection is both necessary and impossible. He capitalizes on this ambiguity in order to complete the system with respect both to the object and to its subjective conditions. Of course, Fichte begins with the explicit recognition that there is no direct access to the generative act that accounts at once for the emergence of consciousness and for its constituted structure. Actual consciousness has always already outpaced its act of birth, it lies "ahead of" this act. But Fichte thinks that this "ahead" must also have an initial moment. This is the moment of every *new* series in consciousness, which is like "awakening from a deep sleep or from the inability to perceive, especially in the case of a place with which we are unfamiliar."[9] We can sense overtones here of Pascal's fright at awakening on an unknown desert island. But Fichte is not content to accept the failure of all attempts to reconcile the experience of self-knowledge and the experience of the world. He has great confidence in the new series, which is supposed to allow us to recover the first series in its entirety. Surprised at awakening in an unknown place, I find myself once again in my own being—the Self. Then I turn my attention to the things that surround me, in order to get oriented. This allows me finally to ask legitimately about the *absolutely* first series: Where am I? how did I get here? what has happened to me?

Consequently, Fichte finds that the sought-for amplification of the mind will succeed if we are ready to think thoroughly in accordance with the "duplicity" of consciousness which characterizes thinking, at least while our primary reflection lasts. Is the reflective judgment incapable of giving itself what will be given in the end? It is sufficient to take reflection as a perspective on this end, which is nothing but experience as a whole. Here, the end justifies the means. The abyss between the a priori and the a posteriori is now reduced to two different perspectives on the absolute totality of things. In fact, reflection can afford to be indifferent with respect to what will emerge in the end, because what is given in the perspective is also given as part of the absolute totality. Thus, we may ask, do we need two points in order to draw a straight line? In Fichte's opinion, if only a point (the starting point of reflection) is given together with an angle, then "all the points on the line are given along with it" even though experience has not yet run through them.[10] When Kant's cosmological principle thus becomes the norm of any possible philosophical activity, the way is clear for rethinking the primordial relation of object to subject. The relation between subject and object can now be considered both *before* and *after* any synthetic activity of consciousness. Before a first synthesis occurs, they are mere opposites, they constitute a pure thought (referred to as the absolute) without the slightest reality. Such a pure thought forms the basis of the first principle of any possible knowledge or being—this is the principle of the absolute self: "The self begins by an absolute positing of its own existence."[11] Kant has taught us that existence is not real predicate. Therefore, existence is spread through the judgment by means of the copula that links the subject to the predicate. The first principle, which is a first reflection that does not yet know itself as reflection, can be stated as an affirmation of existence which has no ground that can be thought or stated in the form of a judgment. Clearly, in order to posit the first principle, Fichte needs to violate Kant's rule for any possible transcendental proof: instead of viewing the reciprocal movement of the ground and what is grounded as one that cannot be overcome, Fichte says that the ground must fall outside what it grounds precisely because the former is supposed to explain the latter.[12]

The positing of the absolute self is merely the coming into being for us of what is formally identical to itself in its own being. But precisely because it is absolute, this form is still nothing determinate for us, even though our consciousness has apprehended it. If we then reflect upon the conditions which have enabled us to posit the first principle, we will oppose a second principle to the first. The second principle thus finds its expression in the form of a contradiction: a not-self is opposed absolutely to the self.[13] Again, nothing is thereby determined for us, because the not-self too is posited absolutely. If the two absolute principles are combined absolutely, they will simply annihilate each other, which is another way of proving that they are still nothing for us. The next reflection focuses on what allows these two opposites to coexist in our own consciousness. This gives rise to the third principle, according to which the self posits itself as limited, and thus *determined* by the not-self. A synthesis of the two principles is thus possible if, instead of mutually destroying each other, they limit each other, that is, if they are divisible so that a certain quantum of self coexists with a certain quantum of not-self. Fichte achieves what was impossible for Kant. Because the movement from one reflection to the next takes place in an absolute totality that is always already given, it gives us an insight into the actual genesis of representation. What we have called above the absolute quantity of pure intuition is now determined in terms of a process constitutive of any consciousness. Fichte characterizes such a process as a *deduction* of what is presupposed in any determination. At the beginning, our consciousness is filled by nothing, except by what Fichte calls a "salutary illusion of the imagination which, without my knowledge, ascribed a substrate to these pure opposites."[14] The transcendental illusion is thus no longer as grave as it was for Kant's notion of reason. To be sure, we cannot begin by reflecting upon nothing, but our deceptive imagination provides us with the elementary material which was needed. Fichte finds that the illusion itself was not to be avoided or eliminated. All we had to eliminate in order for reflection to materialize itself into determination was the product of the illusion. The absolute totality is no longer an Idea of Reason, but something imagined, which Fichte conceives, in the final analysis, as allowing us to deduce the sum total of categories and principles of understanding.

On balance, Fichte attacks transcendental philosophy from above only in order to surpass it. The whole world is within us. The world and the concept of the world are but two perspectives on the same thing. To reach this point, Fichte borrows his model from that of God's thinking: "Our idea of God and God are the same thing, and this idea is objective, that is, it is something from which we cannot separate ourselves and which is as certainly in us as it is certain that we exist."[15]

The Future as a Transcendental Problem

The intellectual journey of Maimon and Fichte reveals that perhaps even Kant could not remain entirely faithful to his own ultimate principle, which states that "all our knowledge relates, finally, to possible intuitions, for it is through them alone that an object is given" (A719/B747). Possible intuitions can be related to a priori concepts in either one of three ways. (1) Mathematics: a pure intuition (such as space or time) is already built into the concept, which can therefore be constructed. (2) Mathematical principles of pure understanding: the concept contains the rule for the synthesis of possible intuitions which are not given a priori, but which can be represented a posteriori (by means of the rule) in perception. (3) Dynamical principles: the rule included in the concept enables only to search empirically for a synthetic unity of that which can never be perceived. We know that the Transcendental Dialectic results from a "natural" amplification of the dynamical principles referred to as the "Analogies of Experience." This amplification is based on the fact that the mathematical principles ("Axioms" and "Anticipations") give us a *complete* manifold of intuition, whereas the "Analogies" define only a principle of search because the experience of the manifold is incomplete. However, if Maimon were right, it would no longer be certain that the "Anticipations" give a complete manifold to be synthesized in some way. Sensation would now be given immediately by the thing in itself, so that the transcendental possibility of its synthesis—and not just its final product—would already require mathematics. If *all* degrees from 0 to X can be anticipated only by including the thing in itself as the matter of sensation, Kant would have come close to contradicting himself when he said, as a general truth of critical philosophy, that "the matter of appearances, by which things are given us in space and time, can only be represented in perception, and therefore a posteriori" (A720/B748). Perhaps, without Kant being able to realize it, the "Anticipations" have already brought us into the dialectical domain of necessary illusion. If it were the case that the "Anticipations" provide us with an incomplete experience, an enlargement of the scope of Kant's Dialectic would be made possible; we would have more than three Ideas of Reason.

Does the Transcendental Dialectic itself give us any clue as to how it might be enlarged? Indeed it does. We can find this clue if we follow the direction given in Fichte's *Science of Knowledge*. Fichte completes Kant's project by reversing the orders of truth and illusion, that is, by taking *explicitly* the Dialectic as the core of critical philosophy—something which, against Heidegger, we have asserted to be only implicit in Kant. The absolute of the first principle is already a necessary illusion, because it can be posited without being anything for us. In other words, Fichte's method might indicate how to exhaust the Dialectic; since the Dialectic is its starting point, in principle, all its consequences can be drawn from it. In fact, the theoretical part of the first *Science of Knowledge* (1794) concludes with the assertion that a reciprocal action exists between the self and the not-self. Neither has the exclusive privilege of being finite or infinite. On the contrary, once either term is infinite, the other is finite. According to Fichte, this is the basis of the antinomies Kant established.[16]

We are brought back to the ultimate horizon of the whole critical enterprise. The cosmic concept of philosophy, that Kant aims to revive in modern times, requires that nature be entirely thought through before reason can concern itself with its original ends: moral life and practical freedom. Nature and freedom: the mastery of both objects is of course unattainable by ordinary mortals. In the very last lines of his *Opus Postumum*, Kant seems to exalt the poetic and mythic figure of Zoroaster as the sole prototype representing the highest end of human life: "Zoroaster: the ideal of physical, and, at the same time, moral-practical reason unified in a single sensible object." For humans in general, philosophy is no more than a perpetually unfulfilled "effort toward [such] wisdom."[17] But might not this effort be reduced to nothing in light of the ultimate ends?

Socrates had turned Protagoras' absolute relativism against him by posing the question of the *future*: while each individual's present situation allows him to make judgments about what is, such is not the case when it comes to what should be. Regarding the future, not just anybody, but only the wise can benefit from the authority to set standards. Only the wise can substitute better opinions for worse. In light of the future, then, it is simply no longer true that what appears to me is what is.[18]

Now, for Kant, any imagination that supposedly intuits the future is simply deluded (B270). Far from allowing us to speculate freely about possible future events (as was the case for Hume), the imagination only suspends the present course of events. It makes us reverse the order of time: either we invert the priority of experience over our capacity to understand, or, retaining this priority, we project the prototype of a "primary experience" far into the past, back to an absolute origin which we presume contains the cause of *all* the phenomenality of experience. From the transcendental point of view, the reversal is always double. Through a mode of consequentiality proper to the categories, the priority of the understanding, which conditions all phenomena, inevitably leads us to posit something absolutely unconditioned that encompasses all conditioned phenomena. In other words, the only way to overcome Hume's skepticism is to show that the future is not at all a transcendental problem, as we have done by following Kant's treatment of the cosmological antinomies of pure reason.

Of course, the future would become a problem for practical reason, because the notion of the sovereign good demands that the reign of nature and the reign of ends be ultimately reconciled. But were we immediately to move on to consider the practical realm, we might leave something unfinished in the consideration of the theoretical. The same circumspection which led us to delay converting the idea of totality into an ontological problem is in order here. Before moving to the practical, we might ask why the future is not even an inevitable illusion of theoretical reason. For Kant, reason does not suppose the question of the totality of consequences of possible conditions, for consequences do not make their conditions possible (A410–11; A331; A336–37). Consequences are an arbitrary and not a necessary problem for pure reason.

Heidegger points out that the rejection of consequences as a rational problem is valid for corporeal nature only, but certainly not for history. For a historical event, understanding it by means of its consequences is essential. This is true for the event at any time: even a past event must be determined by the possibility of its future—this future could be our own present. That is why, Heidegger goes on to argue, "the history of the present is a non-sense."[19] Are we then committed to Heidegger's view, according to which Kant simply ignores this other dimension of being, as if the Kantian field of phenomena were exhausted by the domain of the ready-to-hand? It cannot be denied that Kant might well have good reasons for rejecting the consequences as a rational problem, but he certainly does not fail to take them into consideration: he says explicitly that there is no Copernican revolution in history, because this would require a viewpoint that can both see and predict the free actions of human beings.[20] Unlike the Copernican revolution in astronomy, where a position of rest within one reference frame is always possible, there is no such viewpoint for us in the historical course of human events; we are always sent back to another reference.

History does not justify freedom. But could we not say, conversely, that freedom justifies the very movement of history? Fichte tried to develop such an approach of history in his early reflexions on the French Revolution (1793). In the immediately subsequent *Science of Knowledge*, he goes on to give a theoretical basis to this possibility. In order to understand this, let us go back one more time to Kant's sense of arbitrariness concerning the future.

There seems to be a correspondence between the division of the problem of pure reason into reality (the past) and chimera (the future) and a certain restriction of the problem of reason in Kant. Reason does not produce any concepts; it only frees the already existent concepts of the understanding from the limits imposed by possible experience. It is a matter of achieving the complete comprehensibility of what is given. This implies the given reality of the regressive synthesis, which points towards ever more remote conditions in the past. The progressive synthesis, on the other hand, is only *capable of being given*. The given contains within it a possible totality which can always be unfolded. This means that the present is pregnant with the past since it is already by itself a synthesis. To unfold that which is contained in the synthesis is to retrieve something forgotten: one runs through the synthesis contained in the given in an opposite direction to the one which occasioned it. Thus, what a synthesis is by itself is, for us, to be unfolded as an antithesis.

Fichte gives a purely formal justification for the inseparability of synthesis and antithesis, outside the temporality Kant assigns to regressive synthesis: "There can be no antithesis without a synthesis; for antithesis consists merely in seeking out the point of opposition between things that are alike;

but these like things would not be alike if they had not first been equated in an act of synthesis."[21] Retrieving was has been forgotten does not pull us back towards the past. And yet we are still within the realm of transcendental logic. That is why the form in question here is a very peculiar one. If ordinary concepts are set against one another, we can raise ourselves to a higher concept (one with greater extension) that combines them in a synthesis. Conversely, two identical things seem opposites from the point of view of a lower concept to which one can descend. The self, however, as something merely posited, must contain the absolute totality of all conditions. A not-self can thus be opposed to it only formally. The fact that the two terms are brought together in one sphere does not raise us up to a higher concept, since one of them is already absolutely comprehensive. To reach a synthesis, we have to conceive a series in which the absolute self moves towards a lower concept. To the absolute self, Fichte thus attributes the character of divisibility, one which is more spatial than temporal.[22]

Fichte's general criticism of Kant is that he limited the scope of the transcendental deduction of the categories. Kant's task in the deduction was simply to show that *there are* categories, and not which categories belong specifically to the human mind. Similarly, crossing the boundaries of the understanding, why does reason limit itself to just these concepts? The Idea of Reason seeks to extend the concept of the understanding beyond the bounds of the empirical while retaining a connection with it. At a certain point in the past, when we raise ourselves toward higher conditions, the connection with the empirical is broken for us, though we presume that this connection must continue to exist in itself. Regarding the descent toward consequences, could we not say that the present constitutes the same kind of rupture? Future events would thus constitute a leap into the in-itself, justified by the same concern with continuity. In the regressive synthesis, one cannot necessarily see *at what point* the rupture occurs in the continual slippage of the phenomenal part of the series into its transcendent part. There seems to be just as much arbitrariness in the series of the past (knowing at what point the rupture occurs) as there is in that of the future (knowing whether the series will come to an end). The only necessity that could be found in the future brings us to the *final moment* of history. For such a moment is contradictory, not just arbitrary: it should mark both the end of the sensible world and the beginning of the intelligible world, so that at that moment the two worlds would belong to the same temporal series.[23] But why doesn't the contradiction of the final moment of the sensible world compel reason to an antinomical self-division? Kant seems to argue that I cannot think the future without relating my thought to the image of the present, whereas in the case of the past I can think a past series in such a way that, under the influence of the transcendental illusion, I can push the series further and further without being really compelled (at least as long as the illusion lasts) to return to the present from which I started.

One may transpose the present into the past in order to make the latter seem given, but only to a certain extent, for this transposition is an exercise in pure thinking. Kant says that transcendental illusion results from an unobserved influence of sensibility on the understanding. This means that, without my being aware of it, the present forces me to think of the past as a series of present moments that are identical to one another. If this influence were perceived, the sensible in all its diversity would regain its preeminence. To think of the past series from the standpoint of the present is thus a purely formal exercise which makes all moments homogeneous. Because the present is the only event that is of itself (without the intervention of thought) at once a form and a content, it is just as much an ultimate condition of possibility as the unconditioned sought in the regressive synthesis.

Fichte understood the eminent role of the present in the antithetic. The present pushes the antithetic into another order of time. He writes: "There is no first moment of consciousness; there is only ever a second moment."[24] Our reflection on what has happened in the past must be distinguished from the past that necessarily exists for us. In itself, the past is not. It exists only for a consciousness that is possible in the present, one that reflects on the condition of the present and wonders whether some time has actually passed. Since the past truly exists only in an inquisitive reflection, reflection is not hindered in its freedom by the activity of the prior series. The present is far from being completely weighted down by the past. Rather, Fichte writes, "in it, any other perception could just as well have arisen." The regression towards the past unconditioned retrospectively runs through perceptions that could not have been different than they were. It is made possible by consciousness in its *identity*. This process of retrospection can go on indefinitely, because all differences have been eliminated. But the *freedom* of consciousness is not limited to this identity. Consciousness is still free to direct its reflection to other objects besides the past, in order to recover difference.

Though for Kant any imagination that intuits the future is simply groundless, such is not the case for the intuitive understanding. This understanding constitutes a genuine problem for pure reason. Following Fichte, we could say that Kant's idealism cannot explain the passage from the limited to the unlimited—precisely because the unlimited with which one begins in the *Critique of Pure Reason* (space and time) is a mixture of the human and the superhuman.[25] But the passage from the order of determination to that of *reflection* leads Kant to rediscover the future as a transcendental problem, because implicitly he has returned to Hume's problem. The discussion of the problematic relationship between human understanding and an intuitive understanding is in fact the only place where a leap into the unlimited has a transcendental justification.

Transition to Life

The Transcendental Dialectic of the first *Critique* was Kant's attempt to come to terms with the residual action of sensibility over the understanding, a residue which justifies the transition to ideas. The peculiar relation of ideas to categories shows the internal reasons that explain why the Copernican revolution cannot be completed on its own terms. This residual action has the effect of making the bounds of pure reason burst apart, even though the main business of the critique remains to determine these bounds—namely, to set bounds to reason by surveying the territory within which it can exercise itself in accordance with its own principles. In the *Critique of Judgment*, the explosion of reason's sphere makes the limits themselves burst into pieces. Kant discovers that there are certain kinds of sensible objects which, as particulars, are presented to us in such a way that they completely escape the grip of reason. What are they?

In accordance with the cosmological principle of the first *Critique*, we are invited always to extend the search for a member in the series of possible experience, even when experience has reached its actual limit. In this way, the absolutely unconditioned is never to be met effectively. Now, there is something quite paradoxical, and perhaps even self-defeating in the regressive synthesis (even though it helps solve the antinomies), because reason does not seem to give itself the means for its own ambition. Reason is moved by the desire for the unconditioned which is absolutely *other* than any conditioned. But in order to reach it, reason constitutes a synthesis of ever expanding familiarity, because it is grounded in the connection of appearances as appearances of appearances. There will be members in the series of objects of experience which are so remote from their immediate origin that they could be referred to as purely cosmic objects. Even though they may not be known by experience, they have the peculiarity of allowing to be presented in immediate intuition. There can never be another Newton, Kant argues, "who would explain to us, in terms of natural laws unordered by any intention, how even a mere blade of grass is produced" (CJ 400). As an organized being, the blade of grass is both *the remotest* (because it exhibits a purpose, but "purposes in nature are not given to us by the object: we do not actually *observe* purposes in nature as intentional ones" [CJ 399]), and *the closest* (because within immediate reach of the senses). Thus, in the context of laws of nature which relate to objects such as the blade of grass, Kant indicates explicitly a possible reconciliation between common understanding and philosophy (CJ

398). (It is precisely this accord which will constitute the starting point of moral philosophy.) Indeed, both agree to attribute the production of such an organized thing as the blade of grass to a supreme cause existing *apart* from the world. They differ only in respect of how this place apart from the world is postulated and what it actually explains. The common understanding is dogmatic inasmuch as it assumes the objective character of the proposition that an intelligent original being exists outside of the world of appearences. The critical philosopher acknowledges that the concept of purpose is merely *added* to nature's products "in our *thought*, as a guide for judgment in reflecting on these products" (CJ 399). The critical project concerning organized beings is thus again a project of amplification of thought beyond possible experience. The fundamental difference has to do with the objective correlate of this expansion. An organized being already adds, by itself, something—namely, an organization—to the kind of objects dealt with in ordinary experience. Taken at the level of raw sense impressions originating from objects in general, the manifold of intuition lacks all organization, which in fact is supplied by the concepts of understanding. If there is anything like a concept of purpose, it will therefore add even more organization to an organized being, and it is this doubling which is now under examination. In this process, thought discovers its own subjective character as lying apart from the world (of sense), and the question that we have to address is whether this expansion leads again to a form of illusion which can be discerned in transcendental terms.

All we can assert thus far has to do with the degree of comparison between ordinary and living objects. As far as the former are concerned, the upshot of the transcendental critique is that the allegedly objective correlates of thought withdraw within the subjective sphere. In the case of the latter, the subjective sphere itself withdraws and is no longer part of the world. Since the world is always finite from the standpoint of sense, there is at this point a significant opening toward the infinite. Far from breaking off with the fundamental presuppositions of critical philosophy, Kant is now about to give them their highest scope. He is pointing to the notion that the thought of the finite by finite and actually living beings presupposes or even implies that of the infinite. Understanding in its empirical employment was already busy redoing all that had always and already been done in sensibility, namely, the infinity of its forms. But in order to grasp ourselves as finite, and therefore find out the specific character of our understanding, it is necessary to refer to a supreme cause (a higher intellect) which, even though it is an emanation of ours (since we postulated it), abolishes any distinction between possibility and actuality. The distinction between the possible and the actual corresponds to the requirement of two heterogeneous components in cognitive powers, understanding and sensibility. For the higher intellect hinted at by the concept of purpose, there would be no possibility to be distinguished from necessity, but only one indivisible reality. Such an intellect would thus be pure act, free of all passivity, for which the world, *and* all it contains, would be *immediately* present. Kant says of this intellect that it is "archetypal" (CJ 408). Recall the conclusion which had led Kant to postulate his cosmological principle in order to rescue the idea of the universe from collapsing in the antinomies: we do not know what enables us to assert that the universe continues to possess a magnitude (it can be neither an appearance nor a thing in itself), but the principle gives at least a negative determination to this magnitude, which is the only one possible in view of the form of our inner sense. As a negative determination *within* the world, it is compatible with a positive determination *outside* the world; this latter positivity is the archetypal intellect.

In order to make the production of the blade of grass intelligible, a regulative principle which no longer belongs to reason alone is thus needed: this is the principle of reflective judgment. This principle is the finishing touch to the critical edifice, since it has to combine the remotest and the closest. When we reflect on the objects of nature, Kant says, we do so "with the aim of having thoroughly coherent experience" (CJ 184). What, then, is the principle of reflective judgment? The subjective sphere has gained complete autonomy from the world. Kant calls "heautonomy" a principle of judgment that holds for the subject only, and not also for an object (CJ 185). It will consist in an extension of the cosmological principle, which can be expanded from the *objects* of knowledge to the *principles* of knowledge. The principle can be defined by contrast with another kind of judgment which befalls reason entirely (CJ 211):

> Judgment can be regarded either as merely an ability to *reflect*, in terms of a certain principle, on a given presentation so as to make a concept possible, or as an ability to *determine* an underlying concept by means of a given empirical presentation. In the first case it is the *reflective*, in the second the *determinative power of judgment*. . . . The principle by which we reflect on given objects of nature is this: that for all natural things *concepts* can be found that are determined empirically.

Elsewhere Kant gives this further explanation (CJ 385):

> When judgment *reflects* . . . it has to subsume under a law that is not yet given, and hence must subsume under a law that is in fact only a principle of reflection on certain objects for which we have no objective law at all, no concept of the object adequate as a principle for the cases that occur.

The principle of reflective judgment is thus always to be sought (just like new objects of experience in the cosmological principle), whereas in the case of determinative judgment, the principle is not even to be sought, since it can be identified with the concept or the rule provided by the understanding. Now, what provides the guide for the search of the principle of reflection is not a new object of experience, but rather those maxims of judgment that are found in physics and metaphysics, by which we think the a priori condition under which empirical concepts of objects "can be *further* determined a priori" (CJ 181).[26] What is this further a priori? These are the maxims of judgment at work in already constituted physics and metaphysics, such as the principle of least action (formulated by Maupertuis for the first time in 1748) according to which, if there occurs some change in nature, the amount of action necessary for this change must be as small as possible. It goes further a priori than Newton's laws of motion, because it can be used to derive the equations of motion in other areas of physics—such as optics—once the appropriate action is identified. With principles of this kind, physics and metaphysics take care *by themselves* of what concerns "no more than the possibility of experience and hence of our cognition of nature" (CJ 182); at that level, they have no need for any transcendental critique aimed at grounding their assertions. Yet Kant maintains that the principle of *purposiveness* of nature in the diversity of its empirical laws, as distinct from the mere concept of *purpose*, is a transcendental, not a metaphysical principle. It is thus extracted from already constituted physics and metaphysics, instead of being established in anticipation of any such constitution. This reversal of the Copernican reversal does not annihilate the original reversal, but adds a dimension to the transcendental imposed by the facts of life and organization.

The primary implication of this definition is that reflective judgment takes place spontaneously, without any mediation by determinative conditions, even though it underlies all search whereby, in the end, we expect an agreement between our reflection about nature and nature itself (CJ 212). Because it is heautonomous, giving its own principle to itself, and also because it is subjective without any need (or possibility) for discursive proof in the world of appearances, reflective judgment occurs immediately, outside of time. This immediacy, however, is not identical to the mode in which mathematical objects present themselves after they have been constructed in intuition. This is because the immediacy involved in heautonomy, instead of keeping the present in suspension, as it were, opens up the possibility of thinking the *future* in transcendental terms. There are essentially two instances of reflective judgment, and both of them instantiate this new possibility.

The first is the *judgment of taste*. Its structure is the following: on the basis of my own feeling of pleasure in an object, independently of the object's concept (and thus reflectively), I judge this pleasure as one attaching to the presentation of that same object in all other subjects of the human species (CJ 290). I do not even wait for other people's assent before attributing this feeling to them. A leap is made beyond the limits of experience: I install myself in thought into the mind of another person, without bothering to see whether it can be effectively given in an experience, as was still the case for theoretical knowledge. The judgment of taste is structurally different from the propositions of pure a priori understanding in the following respect: while in the "Anticipations of Perceptions" the manifold of a local experience is speeding up in order to restore the continuity of that experience, and while in the "Analogies of Experience" a manifold that lends itself to an allegedly global experience is always postponed, in the judgment of taste there is no waiting for confirmation of my own feeling to be shared by others, and furthermore, a new sense of globality is brought about, namely, the human species as a totality in its own right. We could thus say that, whereas in theoretical knowledge an illusion remains even after the critique has detected it, in matters of taste each individual asserts his membership to the human species by harbouring the illusion that his feeling will be automatically confirmed by others.

The judgment concerning organized being (such as the living organism) is the other kind of reflective judgment. Kant states that the principle of life has no need at all for any kind of regressive synthesis, because each part of an organism contains the whole.[27] The organic body is by itself, without any imposition on the part of our understanding, a "maximum of progress."[28] That is, it is not different from an idea,[29] inasmuch as its reality is never quite fixed. Since it assembles the remotest and the closest, the whole universe finally appears as a "cosmic body"[30] which is formed organically, not mechanically. (In fact, it is significant to note that the earth can be already regarded as a vast organism.)[31] There is an absolute difference between life and matter, because "an intention can never be a power of matter."[32] Matter is always a composite, the organization of which cannot be drawn from moving forces, whereas an intention is an absolute unity of a subject which combines the manifold of representation.

Whether it can be characterized by the demand of an assent that has yet to be fulfilled (in the case of taste), or whether it is a forward-oriented intention (in the case of life), reflection takes us to the future. But to think the future transcendentally, even though it is not a problem for theoretical reason, requires no less than transcending (and thus reversing) the very Copernican reversal that gave rise to the possibility of transcendental determination in the first place. The extraordinary difficulty involved in exempting reason from determining anything in transcendental terms with regard to the future is pointed out in the "Appendix" to the Transcendantal Dialectic of the first *Critique*. Kant ascribes to the employment of reason a difference which, in the *Critique of Judgment*, will fall within the purview of the power of judgment alone. This difference concerns the nature of the universal (which is used as rule for any subsumption) from which reason derives the particular

(A646/B674). Either this universal is already certain by itself, in which case reason is used apodeictically and the particular is determined in accordance with the necessity of the subsumption. Or it is admitted only problematically (as an idea), in which case reason is used hypothetically; universality can be approached only through the accumulation of particular instances, which are made certain only so far as the rule has reached or will reach. This hypothetical employment of reason is, of course, limited to a regulative function. As Kant puts it: "transcendental ideas never allow of any constitutive employment . . . they have an excellent, and indeed indispensably necessary, regulative employment, namely, that of directing the understanding towards a certain goal upon which the routes marked out by all its rules converge" (A644/B672). This direction consists in a collecting: the ideas bestow a systematic unity upon the concepts of understanding, in accordance with their greatest possible extension. At this point, the triad of scholastic transcendentals (unity, plurality, totality) can be reinterpreted critically, since the ideas give the concepts their greatest possible unity together with their greatest possible extension; the scholastic transcendentals are now grounded in an indubitable material content, namely, the particular instances which are all certain under the hypothetic rule.

But what exactly is this "certain goal" that Kant does not really specify, except in terms of *focus imaginarius*? It seems extraordinarily difficult to ascribe any kind of content to it, even though an experienceable content of some kind is always the touchstone whereby a critical argument asserts itself. The fundamental mistake committed by the Scholastics, as Kant pointed out earlier on (see chapter 3), precisely rested on a confusion between a direction of thought, which is offered by general (merely formal) logic, and the result (or determination) of thought under the form of knowledge. Now, does not the distinction between the two employments of reason (apodeictic and hypothetic) correspond to this distinction made by Kant against the Scholastics? If this were the case, then the hypothetical employment of reason would render the systematic unity of concepts purely subjective in a sense that would not really be different from that of formal logic. At this limiting point, the sharpest critique seems to fall back on crude dogmatism. Does critical philosophy have the resources to struggle against a form of dogmatism which, as it now seems, is its own extrusion—a dogmatism which it has generated and domesticated? Is there anything that we can oppose such dogmatic outgrowth of the critique? The issue is raised with peculiar acuteness in the following passage (CJ 395):

> We treat a concept (even an empirically conditioned one) dogmatically if we consider it as contained under, and determine it in accordance with, another concept of the object such that this other concept amounts to a principle of reason. We treat a concept merely critically if we consider it only in

> relation to our cognitive power, and hence in relation to the subjective conditions under which we think it, without venturing to decide anything about its object.

What, then, if the subjective conditions cannot go beyond the level of formal logic?

To the logical principle of systematic unity Kant opposes what he calls a *transcendental principle* (A648/B676). It does not only guide the search toward the greatest possible unity, but it also postulates a priori that the nature of the object and the nature of the understanding are such that they do lend themselves to systematic unity. It is an *objective* principle which postulates something about the object or the organization of knowledge. To be sure, this principle comes after the cosmological antinomies have already been "solved" critically, by means of the rejection of a transcendental principle stating that reason knows the unconditioned as object. In the new version of the transcendental principle, something is actually decided with regard to the nature of the object, but only inasmuch as this nature cannot be wholly inconsistent with the demand of systematic unity. How is this nature proper to systematic unity? In asking this question, we proceed to a reversal within the overall structure of the critical argument, since, according to Kant, the only way of answering it is by thinking the transcendental principle as a condition of possibility for the logical principle (A651/B679). This amounts to no less that assuming a condition in the object as prior to subjective conditions. Kant's justification of this answer is not easy to follow, but the difficulty has to do with the fact that critical thinking cannot but extrude a form of dogmatism of its own.

The difficulty can be expressed is the following way. Kant had begun the argument in the Dialectic by taking great care in putting out of play a faculty of originary imagination that would wrongfully assume the role of a faculty of reason. Now, by identifying the direction of the understanding with a *focus imaginarius*, he seems to lay an imaginary object at the foundation of the rational idea. Thus, it becomes possible to "derive the object of experience from the supposed object of this idea, viewed as the ground or cause of the object of experience" (A670/B698). The supposition is an act of imagination, in the sense that the things of the world must then be viewed, for example, "*as if* they received their existence from a highest intelligence." What is involved here is the transcendental status of a real connivance or complicity between our thought and the world. Kant does speak of the constitution of the objects *or* the nature of the understanding, and goes on to ask if either of them "is in itself determined to systematic unity" (A648/B676). More generally, our problem now goes something like this. If the regulative principle in the hypothetical employment of reason (which we have called the cosmological principle) has already been instrumental in destroying the constitutive illusion of ideas, will not the above-mentioned reversal, in which imagination assumes a new commanding role over reason's critical claims, compel us to examine the possible dialectical abuses of this very same regulative principle? As an ultimate implication of the argument in favor of the primacy of the objective (as opposed to merely logical) principle, we will, as a matter of fact, come up against the antinomy of teleological judgment, which is of a very special type indeed.

The logical principle of reason is essentially a principle of parsimony, whereby individual things or concepts can always be derived from higher and more general classes. Kant accepts without reservation the scholastic rule, namely, that "principles must not be unnecessarily multiplied" (A652/B680). It is a rule of reason, and yet we cannot help but keep expecting it to be also a rule of things themselves. Significantly enough, Kant's examples in support of this rule are no longer borrowed from modern mathematical physics. They are drawn from chemistry, which was then far from any Copernican revolution in its principles. As a contemporary of Lavoisier, Kant was not in a position to know any chemistry other than Stahl's, which advocates the existence of phlogiston. He cites, for instance, the reduction of all salts to two main genera, acids and alkalies, assuming chemistry to be no more than some kind of sophisticated experimental art. In the *Metaphysical Foundations*, despite his praise for Stahl, Kant claims that chemistry is not a science at all, because it lacks apodeictic foundations.[33] Lavoisier's epoch-making *Traité Elémentaire de Chimie*, in which mathematics is applied to chemical concepts, appeared in 1789. There is no doubt that this made Kant realize that chemistry now referred to a substance similar to that of which physics tries to formulate the laws.[34] But significantly enough, prior to this revolution in chemical science, Kant used the analogy of the transcendental philosopher and the chemist: the transcendental experiment of pure reason, he writes in the preface to the second edition of the *Critique of Pure Reason*, "bears a great similarity to what in chemistry in sometimes entitled the experiment of *reduction*, or more usually the *synthetic* process" (Bxxi, note). This is the process of separation of a given unit into two heterogeneous elements, the noumena and the appearances. If the subsequent transformation of chemistry into a science has any effect on Kant's thought, it lies essentially in refining the analogy whereby the transcendental philosopher distinguishes himself from the mathematical physicist, not in imposing radical (and perhaps unwelcome) changes in the status of transcendental foundation.[35] Indeed, if we are to go even further, if only the one ultimate material substratum or fundamental force is looked up, then *nothing* will ever prevent us from believing that science will succeed in carrying out this search to completion. It is at this point that the hypothetical principle of reason (logical and regulative) converts itself into an apodeictic principle (transcendental and constitutive) (A650/B678). The determining character of the object is now given to assert itself *against* the Copernican reversal. As a matter of fact, the transcendental principle of unity does not involve the ideas, since at that level nature is nothing other that "those objects only which are given to us" (A654/B682).

The above-mentioned conversion of the logical into the transcendental principle has the peculiarity that it establishes judgment as the power that relates both to the remotest and the closest. Indeed, once armed with its transcendental principle, the understanding can seek knowledge both in the direction of generality (the principle of genera) and in the direction of the determinateness of the content (the principle of the manifoldness of species). There are higher concepts only because there are lower concepts—those that come as near as possible to the sensible particular—from which they can be abstracted (A656/B684). In order to achieve this unity—which, as we have seen, is the mark by which the organized being imposes its constraints upon our power of judgment—Kant must pose as a principle that all the manifold genera are simply divisions of one single highest and universal genus. According to this principle, the transition from one species to another does not occur by any sudden leap, "but only through all the smaller degrees of difference that mediate between them" (A659/B687). The continuity of forms is an idea, and no more than that (A661/B689). The principle is thus nothing other than a transposition of the principle of the "Anticipations of Perception" from the object to the concept, whereby concepts themselves (instead of elements of sensation) border upon one another so as to leave no empty intervening space. What is it that justifies such a transposition? The transposition results from the critical resumption of the scholastic triad (from unity to plurality and totality), inverting the order of its terms. Indeed, reason delimits the territory of the understanding (and bestows upon it its proper goal) by progressing from a principle of homogeneity (parsimony of first causes: the totality of the manifold is the highest genus), through a principle of variety (the diversity of effects), on to a principle of affinity (between all members of nature) (A657/B685). The three principles taken together make up the transcendental law of nature's continuity. It is unity (and not totality, as it was the case for the Scholastics), which combines the first two principles. Unity is the level at which the affinity of all *concepts* is sufficient to constrain each species of *nature* to follow a gradual increase in diversity. This inversion of order is the price one pays for maintaining the scholastic triad, and adding to it an actual relation to the nature of the object.

But this is an enormous price. A fundamental presupposition of the original Copernican revolution has been omitted. This is nothing other than the movement of the world's spectator. Kant represents systematic unity by means of a static analogy (A658/B686). Every concept is a point which is like a viewpoint: it has its own horizon, within which a variety of things (in fact, every species) can be surveyed. By virtue of continuity, there are an infinite number of points within each such horizon, each of which has its own, narrower, horizon. The point/concept which, ideally, possesses no extent (that is, the individual) is never to be met with in person: the concept cannot but *give to see* something outside of itself. This is true even where there is nothing to see, because "the species in nature are actually divided" (A661/B689): there is, as a matter of fact, no true infinity of intermediate members between any two given species. But over and above this matter of fact, we can postulate the existence of a horizon common to all genera; this horizon has its own center, so that "we arrive at the . . . universal and true horizon, which is determined from the standpoint of the highest concept, and which comprehends under itself all manifoldness" (A659/B687). No displacement of the central point is needed in order to reach this global perspective; all we need to postulate is some kind of communication between many such points. At the limit, these are the many concepts of one and the same thinking subject, so that subjectivity finally constitutes a world of its own, apart from the world of nature. In this separate world, "there are not different, original, first genera, which are isolated from one another." Any radical heterogeneity between its constituents (by way of, say, a separation of intuition from thought) has thus been transcended.

The rejection of separation implies a return to the tree of knowledge: all species spring from the same stem. It also implies the rejection of antinomies. For instance, in order to avoid being lost in the maze of species and genera, Kant says that we are merely "bidden," but not logically compelled, to pay due regard to homogeneity. How is this possible? Kant's transcendental principle is both regulative and constitutive at the same time. It is transcendental, because it regulates the regulative function of reason in its logical employment. But it is also constitutive, because otherwise the understanding would be led astray, it would "follow a path which is perhaps quite contrary to that prescribed by nature itself" (A660/B688). The antinomies are done away with, but since the impossibility of an overlap between appearances and noumena must remain (i.e., the continuity of the forms of nature is only an idea), this "objective" regulative principle of the "subjective" regulative principle is a cosmological principle of an entirely new kind. Indeed, "we could not make any determinate empirical use" of the law of continuity, "since it instructs us only in quite general terms that we are to seek for grades of affinity, and yields no criterion whatsoever as to how far, and in what manner, we are to prosecute the search of them" (A661/B689). That is, the law is *not* a law of existence; it provides us with a sign thanks to which a higher member in the series of experience must be sought, but it furnishes no certainty that this member will ever be found. Limits are now set in advance to the amplification of the "Analogies of Experience"; it was the ignorance of those limits that had led reason to fall into the trap of antinomies. Combining these two notions of relentless search and absence of certainty, Kant declares that all transcendental principles, inasmuch as they are conditions of possibility of logical principles, have an *objective but indeterminate validity* (A663/B691). This is the point at which thought left to itself (without the help of the idea of God) has come as near as possible to the pure, original forms of sensibility, since the latter manifest themselves as forms when the appearance is regarded as undetermined object (A20/B34).

The basic indeterminacy can be expressed as follows. Reason gives systematicity to the manifold acts of the understanding by connecting them, just as the understanding's task is to connect the manifold of appearances in the first place. But there is a difference: we have schemata of sensibility, without which the acts of the understanding would remain undetermined, but there are no schemata of reason (or, if there are any, they come to nothing), which is why the unity of reason *remains* undetermined. However, Kant does not leave the matter at that. Just as the return to unity from the totality was, for the Scholastics, an indication of the perfection of knowledge, so a return to the totality from the unity marks a culminating point for theoretical knowledge in terms of critical philosophy.

> Although we are unable to find in *intuition* a schema for the complete systematic unity of all concepts of the understanding, an *analogon* of such a schema must necessarily allow of being given. This analogon is the idea of the *maximum* in the division and unification of the knowledge of the understanding under one principle. (A665/B693)

The totality thus operates like a schema of unity. And the idea of maximum, Kant adds, is something that "can be determinately thought." We could thus say that the regulative principle of the regulative principle makes determinate what was indeterminate in the original regulative principle. As long as we maintained ourselves within the Copernican revolution, the schema of reason came to nothing. It was pure indeterminateness, because it was the representation of certain methods aimed at sustaining unity in the manifold possible acts of the understanding. Each of the latter is always an experience to do, which must be connected with those that have already been done. But now we find that totality, in combining the remotest and the closest by means of a necessary analogy, provides reason with determinateness. That is, the unification of the acts of understanding is *at the same time* the unification of nature.

No less that such a reversal of the hitherto adopted strategy in the *Critique* is needed in order to proceed to a transcendental deduction of the ideas of reason, a deduction which was rejected in the Transcendental Dialectic itself. Any deduction consists in explaining and grounding a priori the relationship of a concept to its object. Since, in the present case, the object is no more than a schema, the nature of the relationship between concept and object is a modified version of what it was in the Transcendental Analytic (A670–71/B698–99). The result of the deduction of the categories was that the conditions of possibility of experience were equivalent to the conditions of possibility of the objects of experience. By contrast, the result of the deduction of the ideas is that the conditions of the object in the idea are also the conditions of the idea itself. But since the object is a schema, it is also nothing more than a rule needed in order to find *other* objects. Given the absence of certainty which surrounds this search, we find, at the limit, the supposed or imaginary object which operates as ultimate ground of the object of experience. Take the idea of a highest intelligence as cause of the world, an idea which is external to the world of sense, and which, furthermore, is not employed absolutely to determine anything in the object. Nevertheless, I cannot but think this being by analogy with the categories which prevail in the world of sense, such as substance, causality, and necessity. These concepts, Kant argues, "lose all meaning, and are empty titles for possible concepts, themselves entirely without content, when we thus venture with them outside the field of the senses" (A679/B707), something which we cannot stop ourselves from doing. The reversal of the reversal has now come full circle. At the constitutive level of knowledge, intuitions without concepts were senseless, whereas here it is the concept without content which is declared to lose any meaning.[36] Still at the constitutive level, the antinomies gave meaning to the ideas, since the ideas *always* had a content, namely, the transcendental appearances.

The Antinomy of Life: The Often Foolish Reason

Acknowledging the meaninglessness of the cosmological idea with respect to its supersensible object is a turning point in the *Critique of Pure Reason*, since Kant decides to abandon the cosmological idea altogether. The psychological idea and the theological idea are not exposed to an antinomy, and therefore, in the case of these two ideas (but not the cosmological idea) "there is nothing whatsoever to hinder us from *assuming* these ideas to be also objective, that is, from hypostatising them" (A673/B701). As a cosmic concept, that is, as absolute totality in the synthesis of appearances (or concept of the world-whole as *Begriff des Weltganzen*) (A407–8/B434), the cosmological idea cannot be hypostatized, and therefore it would seem that its meaning cannot be rescued from complete collapse.

The possible abuses of the regulative principle concerning its employment with respect to given objects do not lead to antinomies, because an objective (transcendental) principle has been formulated, which has the double (regulative and constitutive) function. Accordingly, the regulative idea of world in general dissolves completely, and gives way to that of *nature* which "is properly the only given object in regard to which reason requires regulative principles" (A684/B712). The world, Kant says, signifies "the mathematical sum-total of all appearances," whereas nature means the world viewed as "a dynamical whole," which concerns "the unity in the *existence* of appearances" (A418–19/B446–47). In the mathematical antinomies, as we recall, both the thesis and the antithesis were declared to be false. As regards nature, the touchstone of the reality of the concept is no longer the empirical intuition of appearances, but rather the existence of such appearances and their relation to one another in respect of their existence. (Existence is taken in the sense given to it in the "Analogies of Experiences" at A178/B220.) In order to apply categories to nature, and determine the latter in its internal possibility, only sensible intuition is required as guide. At the limit to which reason presses the understanding, when categories relate to nothing that can be represented in sensibility (and therefore fail to be actually applied to anything), the idea of absolute totality can still be used with a view to explaining appearances. In that case, (dynamic) nature reduces to (static) world, and we proceed in accordance with the cosmological principle as if the series of appearances were indefinite. But the authentically dynamic character of nature, as opposed to the static world to which it can be reduced (thereby giving rise to pairs of false propositions), manifests itself in that it can also reverse the order of causality. Indeed, in order to fulfill its actual destination, nature can be applied to categories, instead of the other way around. This occurs when "reason is itself regarded as the determining cause"; then "we have to proceed as if we had before us an object, not of the senses, but of the pure understanding" (A685/B713). Such a causality is the sphere of freedom, where an intelligible cause becomes a nature in its own right; it requires practical principles which take us completely outside the series of appearances. These principles are first formulated in the third cosmological antinomy, where it becomes legitimate to conceive a pair of propositions which are both true. Does that mean that the cosmic concept of world wears away once and for all? We are then left to wonder about the occurrence of the antinomy of teleological judgment, where Kant seems once again to reduce nature to world by defining the latter as "the whole of nature" (CJ 414).[37] Unless we want to accuse Kant of redundancy, and claim that the antinomy of teleological judgment (which concerns mechanism versus finality) repeats in its own terms what is already developed in the third cosmological antinomy concerning natural necessity and freedom,[38] we need to develop further the problem of life in so far as it is associated with that of finality.

Life gives a completely new impetus to the whole question concerning the relationship of causality to totality. Life does not mean unity. A living organism is defined essentially as totality in the sense that it is "both cause and effect of itself" (CJ 370). For example, a tree can be said to produce itself in three ways: (i) it generates other trees which belong to the same *species*; (ii) it produces itself as an *individual*, in the way it absorbs and processes the ingredients it receives from nature outside it; (iii) a *part* of a tree also produces itself, because "in one and the same tree we may regard each branch or leaf as merely set into or grafted onto it, and hence as an independent tree that only attaches itself to another one and nourishes itself parasitically" (CJ 371–72). A living organism thus seems to exhibit ostensibly the threefold structure of the transcendental principle (homogeneity-variety-affinity) which regulates the regulative function of our own reason. Totality, however, is no longer merely a schema, so that, instead of withdrawing within itself, the world of pure concept faces again the world of sense.

What is the nature of the series of conditions which pertain to totality as a fact of life? Kant distinguishes between two senses of purposiveness (or lawfulness of the contingent) in nature, relative (or extrinsic) versus intrinsic. If the effect produced by a cause is a purpose, then the purposiveness of the cause is intrinsic to the natural being. If, on the other hand, the effect is itself part of a larger series, so that it in turn serves as means that other causes employ purposively, the purposiveness is merely relative (to these other causes). Consider any series of connected appearances in the natural world around us, for instance rivers which carry along all sorts of soil on which plants can grow (CJ 367–69). The question arises whether the various processes and interactions that can or cannot take place within this series must be judged to be purposes of nature. The answer is yes, because certain natural things pertaining to the series always benefit other creatures as a means, else there would be no life at all. Relative purposiveness can thus be claimed for any such series, in the sense that "in the series of mutually subordinated links in a connection of purposes, each intermediate link must be regarded as a purpose . . . , and its proximate cause is the means to it." In other words, the absolute teleological judgment cannot be justified, because only *one* individual element can be regarded intrinsically, which is totally artificial anyway because the intrinsicality of this single component vanishes when it is included in the larger series to which it necessarily belongs. But, as we know from the cosmological idea, a first unconditioned element in a series of conditioned appearances is only one of the two senses in which we can speak of the unconditioned. The other sense is the complete series. Shall we say, then, that the *complete* series of all natural things, as one single cosmic whole, can be regarded as intrinsic, if at least a privileged vantage point can be found from which the totality can be assigned? As it turns out, there is such a vantage point: this is man, because of the freedom he has in his causality, which prompts him to consider all natural things as beneficial to him.

This relation to nature calls to mind the Ptolemaic standpoint, in which the coincidence of the central point of the universe and man's own position reflects a conviction that the whole of nature is immediately directed to man's uses. Instead of opposing a Copernican readjustment to this view, Kant goes on to argue that the very idea of relative purpose cannot even be assumed. In order to think such a relationship of causality to totality, the field of human reason must be enlarged: "For man's own reason knows how to make things harmonize with the notions that were his own choice, notions to which even nature did not predestine him." When everything is regarded as beneficial to him, man invents, apart from reasonable aims, all sorts of often foolish and dispensable aims (*oft törichten Absichten*), such as plant juices for makeup, or colorful bird feathers to adorn his clothes. The concept of totality associated with life is thus of a special kind: it is neither the sum total of appearances (converted into purposes), nor the absolute and indivisible individual which an organized being would be if withdrawn from the series to which it belongs. Because totality cannot be reduced to either of these senses, dogmatic purposiveness is not, however, to be rejected out of hand. There are, Kant says, hypothetical indications to natural purposes, and, in fact, the concept of a thing as a natural purpose is inevitably dogmatic (CJ 396).

From that point of view, it is perfectly legitimate to lay a transcendental principle of organization at the basis of any logical principle. Indeed, we come closer to characterizing totality in its authentic sense if we note that a living being must constantly, from the very start, *solve a problem*. The problem that the organism must solve has to do with the fact that it embodies the dynamic character of nature at all times; that is, it either perfects itself or tries to retrieve a balance, following the changing requirements of the environment. An organism is an indeterminacy which has the peculiarity of constantly renewing itself in the very act of determining itself. This follows from the special character of the series of conditions: each member of the series must be "a purpose of nature *in its own right*" (CJ 368).[39] Even though it belongs to a species, an organism also retains an individuality. Theoretical reason is unable to constitute a regressive synthesis here, since there can be no appearance of appearance, that is, *there is no concept that corresponds to the purpose of a purpose*. To put it as generally as possible: "strictly speaking . . . the organization of nature has nothing analogous to any causality known to us" (CJ 375). In support of this view, Kant argues that this organization differs from a merely mechanical organization (such as the watch) in the following respects. In the case of the watch, the correlation between parts is achieved by an outsider; the subordination of parts is such that one part cannot produce another; and as far as reproduction goes, a watch, obviously enough, cannot mend itself. The decisive difference has to do with the second of these three points. In the organism, one part produces another, that is, there *is* a series of appearances as appearances of appearances, but it is generated by the organism itself, not by the unifying concept imposed by the requirements of knowledge. When the knowing subject produces the series by constituting appearances of appearances, it can do so only by repeating the same, which is why it can never make intelligible an element of the series which is the limit of that series. But the organization of the organism is internal to itself, in so far as living matter is an example of breakdown in the repetition of the same.

Each member of the series is an individuality that is never completely reducible to others, and yet is in permanent relation to others. Because no sense of totality in terms of the cosmological idea can be ascribed to the natural product, "the concept of natural purposes leads reason into an order of things that is wholly different from that of a mere natural mechanism" (CJ 377). It is essentially by looking at the other two concepts, unity and plurality, that the adequate qualification can be found for totality. A given natural product is a particular form in the combination of matter. If its causality cannot be merely mechanical, it must be based on an idea. But then, if we remain at the level of the concepts of natural mechanism, a conflict occurs between form and matter which can never be bridged. An idea is characterized by absolute unity of presentation, whereas matter is always a plurality of things which cannot generate, by itself, any determinate unity. The only way out of this quandary is to say that "the purpose (the idea) of nature has to be extended to *everything* that is in the product of nature. For once we take such an effect *as a whole* beyond the blind mechanism of nature and refer it to a supersensible basis as determining it, then we must also judge this effect wholly in terms of that principle." Thus, the notion of totality appropriate to natural products is one that applies to an individual member in the series of such products, not the complete series. It is only if the causality responsible for the product's particular form is judged in accordance with the blind mechanism of nature that this product loses any possible link to other such products. What still links it to others, even though it is a totality in its own right, is not the temporal series of the regressive synthesis, but rather a leap: the causality of an idea which goes immediately beyond the blind mechanism of nature.

On the one hand, the self-division of totality cannot prevail here, which seems to protect life from any antinomy of cosmological type. But on the other hand, the foolish dream of reason, which sees itself as the ultimate destination of nature, hurls it down to an antinomy of an entirely new kind. That is why, contrary to Hegel's opinion, it is impossible to reduce this antinomy to the cosmological opposition between determinism and freedom. Kant presents the antinomy in the following way (CJ 387):

> *Thesis*: All production of material things is possible in terms of merely mechanical laws.
>
> *Antithesis*: Some production of material things is not possible in terms of merely mechanical laws.

Three features of this antinomy must be pointed out. (i) Since the idea is not associated with a regressive synthesis, neither the thesis nor the antithesis requires any proof; it is the absence of proof which makes us enter into the above-mentioned "order of things that is wholly different" from that of rational cosmology. (ii) However, the antinomy speaks about material things and their production, and no longer specifically about organized beings; this feature seems to take us back to rational cosmology. (iii) Comparing rational cosmology and teleological judgment, we find out the existence of two dialectical movements, each the inverse of the other. In the case of the mathematical antinomies, each of the conflicting propositions represents an apparently valid attempt at constitutive knowledge, and their opposition is then resolved by collapsing both of them into a single regulative maxim. Not so in the case of teleological judgment, in which Kant has in fact *preceded* the presentation of the antinomy by stating *two* maxims:

> *The first maxim* of judgment is this *thesis*: All production of material things and their forms must be judged to be possible in terms of merely mechanical laws.
>
> *The second maxim* is this *antithesis*: Some products of material nature cannot be judged to be possible in terms of merely mechanical laws. (Judging them requires a quite different causal law—viz., that of final causes.)

The transition from the maxims to the antinomy is realized by means of a conversion, whereby what is assumed to be valid for our investigation of nature becomes constitutive for the possibility of objects themselves. In short, the transition is from the judgment of things to the things themselves. What is really baffling is that both the maxims and the antinomy proper seem to represent an obvious step backwards, compared to the terms of the Analytic of Teleological Judgment. Had we not already concluded that the only viable concept of totality requires the whole of a natural product to be conceived as being based on an idea, not a mechanical necessity? Why does Kant now refrain from using this concept, and why does he distinguish "some" products from "all" products, even though the idea's absolute unity has not really been put into question? Let us try, at least tentatively, to preserve a minimum of symmetry with rational cosmology. In order to do that, we seem compelled to present quite a paradoxical argument, which goes like this: the maxims represent an apparently valid attempt at extending this knowledge of totality beyond some limit, whereas the antinomy itself corrects this attempt by pointing out the existence of the limit, that is, a contradiction that cannot be really surmounted if the limit is overstepped.

Where does the foregoing hypothesis lead us? Consider the first point, namely, the difficulty involved in accepting an antinomy without proof. Shall we say that there are no proofs because thesis and antithesis have already been demonstrated elsewhere?[40] Thus, it would seem that the explanation in accordance with mechanical laws is already demonstrated ostensibly in the Transcendental Analytic of the *Critique of Pure Reason*. Concerning the mechanical mode of explanation, Kant says that "all events in the sensible world stand in thoroughgoing connection in accordance with unchangeable laws is an established principle of the Transcendental Analytic, and allows of no exception" (A536/B564); the third cosmological antinomy makes the same point, by using an apagogic proof. Under these conditions, why does Kant transform the constitutive principle of mechanism into a maxim of reason? A first clue is provided by the fact that he now has to speak about the production, not the connection, of material things. Furthermore, reason comes into play in the second maxim only, which deals with "special experiences," whereas the first maxim is given a priori by the understanding as it deals with the unity of nature in terms of empirical laws. If our hypothesis is correct, then, presumably, the reference to "some products" in the second maxim is an addition to, not a part of, the totality referred to in the first maxim. The transition from the first to the second maxim is thus the attempt, made by reflective judgment, to extend the realm of reflection beyond ordinary experiences—which can be so called, in order to contrast them with "special" experiences.

In an important comment which follows the solution of the antinomy of taste, in the part of the *Critique of Judgment* dealing with aesthetic judgment, Kant takes the trouble to argue, in more detail than anywhere else, that there are no more and no less than three kinds of antinomy of pure reason (CJ 344–45). Strikingly enough, the antinomy of teleological judgment is not counted among them. There are three kinds of antinomy because there are three cognitive powers: understanding, judgment, and reason. Each results from reason's demand for the unconditioned for the given conditioned, whereby this faculty seeks to provide the reason for the a priori principles of each of these higher cognitive powers. The principles themselves are not exposed to any conflict; the conflict arises essentially from their relations to the object, which become problematical when they are posited by reason. In the case of understanding, the extension up to the unconditioned leads to the cosmological antinomies; in the case of judgment, the antinomy concerns its aesthetic use in the feeling of pleasure or displeasure; in the case of reason itself, we could say that reason uses itself to surpass itself and reach the unconditioned, so that an antinomy concerns this double use of reason by reason, that is, the power of desire in the practical use of intrinsically legislative reason. The conspicuous absence of the antinomy of teleological judgment from this list should not lead us to believe that it is redundant with either one of them. In fact, it embraces them all, because it is not strictly limited to the realm of pure reason. Only the antinomies of pure reason care to transform the relation of reason to totality into something valid unconditionally as well. Nowhere does the antinomy of teleological judgment proceed to ascribing unconditionality to totality. That is, it leaves totality as a problem (and life *is* defined as a problem), and that is precisely why the judgments about it become maxims.

There is no demand for the unconditioned in the teleological judgment, because the unconditioned is already contained in its heautonomy, which is *its* a priori principle. Indeed, the faculty of teleological judgment is reflective, without principle, but it tends to procure a coherent subjective view of reality; that is why it serves itself as principle. That the demand for the unconditioned is satisfied in the very principle of judgment is clear from Kant's examination of how we could provide a causal account for the parts of, say, an animal's body (CJ 377). Clearly enough, these parts (the skin, bone, hair, etc.) are always liable to explanation in terms of mechanical laws; there is a priori no limit to our accumulating purely mechanical representations of such parts. But at any point of the series, as we proceed from the conditioned to its condition, we pass from this mechanical type of explanation to another type, the purposive, which is such that *all* the parts are organized according to *one* common cause.

The inclusion of the unconditioned in the principle of reflective judgment brings about unexpected consequences. Since the principle of reflection is not an objective law at all, there is "no concept of the object (which is) adequate as a principle for the cases that occur. . . . Since this principle is not objective and cannot provide an objectively adequate basis for cognizing the object, it has to serve as a merely subjective principle governing the purposive use of our cognitive powers—i.e., our reflection on a (certain) kind of object" (CJ 385). Being absorbed in the principle of judgment, the unconditioned is now the condition of nothing in particular, that is, nothing that is determinate or that can be determined. The maxims of reflection being regulative of themselves, it would seem that reflection is equivalent to nothing other than the purely logical (subjective) employment of reason. In accordance with the old dogmatic dream, could not the logical rules of reasoning be used as maxims for the intelligibility of the world? Perhaps reflection does no more than dress up this logical employment so that it may pass off as being grounded in a transcendental principle. Not surprisingly, then, no less than what Kant refers to as "special experiences" is needed in order to awaken reason from falling into the vacuity of its purely logical employment.

The pair of above-mentioned maxims represents an exposition of just those rules of reasoning, which present themselves as maxims as long as they remain within the realm of reflection. The heart of the argument, however, is that this identification is exposed to danger. In the maxims, the unconditioned remains trapped in the subjective sphere; it is incapable of finding an outlet in the possibility of objects themselves. Why, then, do we not have one maxim which would simply express all possible cases of judgment, instead of two? Remember that these are maxims of judgment, not reason. If these were maxims of reason, the principle of parsimony would immediately come into effect, that is, the principle according to which "reason resists . . . with its maxims to avoid, whenever possible, unnecessary multiplication of principles" (CJ 348). The fact that such multiplication seems to prevail here, indicates that we are already beyond the purely logical (subjective) level. How do we actually go beyond mere logic?

Kant writes that these two maxims "do not seem quite compatible," with the consequence that this "leaves judgment perplexed as to what principle it should follow in its reflection" (CJ 386–87). At this level, the contradiction is only apparent; Kant speaks of a seeming incompatibility or conflict, not an inevitable contradiction. It can be avoided easily enough by paying attention to the double character of the unconditioned which is already included in any judgment related to the forms of natural products. The maxims remain as closely attached as possible to the problem that a natural form in general is; one principle underlying all possible judgment is thus not ruled out. For what the comparison between the two propositions indicates, is that we have to leave "it undecided whether in the inner basis of nature itself, which we do not know, the physical-mechanical connection and the connection in terms of purposes may not, in the same things, be linked in one principle" (CJ 388). Since it does not relate objectively to its objective correlates in nature, not even by way of surreptitious illusion (as would be the case for reason), reflection is at variance with reason, and so it cannot but mistake the two kinds of connection (mechanical and purposive) for two kinds of things. We do not make one maxim compatible with the other by saying that one mode of judgment is not exclusive of the other, as for instance when, in the first maxim, we acknowledge that we did not mean to assert that natural products "*are possible* in terms of mechanical laws *alone*." The power of logic gives out at this very point. The two maxims which serve as principles are not opposed as contradictory; all we can say with a modicum of certainty is that "though *disparate*, [they] might well still be reconcilable" (CJ 391). If, in fact, the reconciliation of those two maxims turned out to involve an impossibility, this is not so much because they contradict one another in a manner which can be logically ascertained, but because reflection never knows unambiguously what object it is dealing with. Thus, the second maxim refers to "certain occasions," it shows that reflection uses the principle of final causes only when it deals "with certain natural forms."

A properly inevitable contradiction between two propositions, that is, the antinomy proper, appears at the next stage only. It results from the conversion of the regulative into the constitutive. Inevitability, which plagues all antinomies of pure reason, comes to its fullest expression in this particular antinomy, since Kant does not really go beyond a "preliminary to the solution of the above antinomy" (title of section 71). In fact, the entire remainder of the *Critique of Judgment* after this section develops the prolegomena to a solution—precisely because no more than prolegomena can be thought in this context. The conflict is not to be solved, since in fact there are no proofs which can be examined critically. It is not correct to argue that we shall be done with the antinomy once the constitutive is again converted into the regulative. Since its exposition comes after the maxims, all we can do is explore the motive for the conversion.[41]

The transition to the constitutive level of determination (legislated by reason) compels us to think in terms of definite totality, where, according to the principle of judgment, there can be no such totality. Armed with its idea of totality, reason asks: What is the true explanation regarding one and the same product and its possibility? That is, reason asks for reconciliation because it mistakes a seeming conflict (that of judgment) for a true conflict. This amounts to misrepresenting reflection, in which the demand of unconditionality has always already been satisfied. The existence of the antinomy reveals that "we still cannot reconcile the two principles *in an explanation* of the same natural product" (CJ 413). When all is said and done, all we shall be able to do, in order to neutralize the antinomical conflict between objective principles of determinative judgment, is to stop "being troubled by the seeming conflict that arises between the two principles for judging" a given product. How can we reach such a position of detachment from conflict?

A Solution of the Antinomy, or the Endless Prolegomena

The antinomy of teleological judgment is certainly the most elusive of all antinomies, and perhaps it hardly deserves to bear the name: there is simply no such antinomy explicitly stated anywhere. For if the conflict between maxims is only a seeming conflict, the conversion of regulative into constitutive principles results in an antinomy which is not even one of judgment, "but instead would be a conflict within the legislation of reason" (CJ 387).

This antinomy is thus the only one in which reason is in conflict with its own principle; that is why it does not even try (because it cannot) prove either of the two propositions. Reason asks for an unconditioned, even though this demand has already been satisfied subjectively by the heautonomy of reflection. This prompts reason to reassert the primacy of its purely subjective basis over objective correlates. Because it reasserts what it already has, reason is now set to appropriate something like a "doubled" unconditioned, that is, a ground that conceals the key to what makes the unconditioned what it is as unconditioned. Such an ultimate ground is nothing other than the intuitive understanding, that is, an understanding which is not discursive like ours. From the standpoint of the intuitive understanding, objects which are *given* to it are not also *produced* independently of it. Kant's "solution" of the antinomy is entirely based on this apparent equivalence between a human reason at variance with itself and an understanding wholly other than ours. Retrieving the theoretical legislation of reason can be achieved only at that price. The leap into the intuitive understanding will not lead us to metaphysical dogmatism; rather, it takes us straight into the "blind spot" of the entire critical enterprise.[42]

The path that will ultimately lead to a solution begins with an argument aimed at dispelling "all semblance of an antinomy between the [two] maxims" (CJ 389). We are still at the level of the merely seeming conflict, that which is not absolutely inevitable. The argument goes like this. When we are interested in grasping the *production* of organized natural products, we collide with a principle which, ultimately, lies in the supersensible. We cannot prove that such products cannot originate from mechanical laws alone, because the first inner basis of nature cannot be reached at all; but, at the same time, our cognitive power does not have the knowledge of mechanics sufficient to account for such production. Hence, in order to be on a par with the sensible, an account of *connection* has to be substituted for an account of production. The following unique principle emerges, which is valid for reflective judgment alone:

> that (to account) for the very manifest connection of things in terms of final causes we must think a causality distinct from mechanism—viz., the causality of an (intelligent) world cause that acts according to purposes.

Thanks to the substitution of connection for production, the two original maxims have been collapsed into one. Unlike its two predecessors, this single maxim does *not* lend itself to further manipulation by determinative judgment, whereby the maxim would be compelled to speak about the inner possibility of objects. Rather, the principle is valid, "however rash and unprovable it would be for determinative judgment." Armed with the foregoing substitution that prevents it from falling back into an antinomy, the maxim seems to side with the antithesis, except that it now speaks about *all* things of nature instead of *some* things—the reference to all events was the language of the thesis prior to the substitution. Indeed, the mechanical explanation loses its original universality and becomes particular, because it is "a guide for reflection, which meanwhile continues to remain open to the discovery of any basis for a mechanical explanation." It is precisely this exchange of "all" and "some" that allows the maxim never to stray from the world of sense, even though it is ultimately conditioned by a supersensible principle into which we have absolutely no insight. To be sure, the exchange cannot be justified on a priori grounds. But at the very least, it is only fair to the fact that the totality, as regards organized beings, cannot be likened to the unconditioned.

If we should still wish to make the new principle valid for determinative judgment, we would demand an unconditioned where, in the world of sense, we already have what Kant refers to as the very manifest connection of things in terms of final causes. (An example of "mere apprehension" which is enough for purposiveness to manifest itself is the case of bodies of regular shape, CJ 411.) No antinomy follows from this additional demand, but only the *probability* that we are misled when trying to reach in the transcendent what is evidently available within the world of sense.[43] The apparent certainty of the antinomy does not obtain here, because sensible purposiveness is already a proof by itself, and we do not know what could possibly be a proof concerning supersensible purposiveness. Therefore, there is *nothing dogmatic* in this shift to the transcendent. However autonomous it may be, the effort of reflection on its proper object (which is the totality, not the unconditioned, because the principle of reflection itself is the unconditioned) is hampered by the very nature of this object, which always remains problematic.

Following this preliminary substitution, which attempts to reconnect judgment with the world of sense, Kant is in a position to launch into the properly inevitable antinomy, which is concomitant with a dogmatic shift to the transcendent. Only one more step is needed in order to draw an antinomy, and not merely a probability, from the conversion of the above-mentioned single maxim for reflective judgment into an objective principle of nature. This consists in admitting that the evidence of the principle of final causes is so massive, already within the world of sense, that it does not even need to be supported by further investigation of nature. The maxim of reflective judgment slips into an antinomy when reason sees in it, in a manner which is precisely unreflective and indemonstrable, the ultimate proof of nature. For the concept of natural purposes, Kant argues, is in fact a "stranger in natural science." Let us, then, "stop our investigation of nature . . . or at least suspend it for a while" (CJ 390). In the entire critical enterprise, this is the point at which the model of natural science is firmly and explicitly put out of play. If we adhered strictly to the bounds imposed by mere cognition of nature, the question as to whether the undisputed maxim is also objectively valid would be left undecided and unsolved.

The evolution of Kant's doctrine is really impressive. In the science of organized beings, we begin by doing away with a merely seeming conflict between maxims of reflection; this preparation allows us firmly to grasp another original antinomy, radically alien to natural science, through which we could perhaps reach beyond nature. The extraordinary difficulty with the latter antinomy which we are about to articulate—compared to the antinomy of pure reason—is that a dogmatic dispute will be followed by a fundamental uncertainty, an uncertainty from which the application of the critical argument was supposed to protect us. There is something deeply unsatisfactory about any scientific theory which dares to explain natural products in terms of final causes. Such an explanation lays claim to the things themselves complying with purposes, for the sake of determinative judgment, while in fact "all we demand . . . is that natural production be judged in a way commensurate with our ability for judging such production, i.e. . . . with reflective judgment" (CJ 408). Even if natural science had complied thoroughly with the bounds imposed on reason by the critique, that is, even if it had refrained from being dragged by an objective purpose before reflecting upon the subjective conditions of any purposiveness, it would still miss the fundamental point about trying to make natural purposiveness intelligible: that we involve ourselves. Our own existence and destination are implied in this involvement, so much so that we find in reason enough resources to transcend it (and thus also nature), without ever violating the critical limits of the sensible world. Through the question of natural purposiveness, we are measured directly against the cosmos, because the preparation by way of the Copernican revolution is henceforth no more than just that—a preparation, not a comprehension. This is a properly abysmal affair, since nothing (neither determinative thought nor the things themselves) indicates a natural limit which we could rely on. In the case of theoretical knowledge the heavens still offered us natural guidelines that the most (formally) general and the most (materially) global perspectives were both compelled to follow. (Recall the intimacy that the cosmological principle maintains with a determinate representation of the cosmos.) It then remained possible to delegate our central position to another point from which the determinate structure of things could more completely be explicated. In theoretical knowledge, reason reasons hypothetically because in accordance with the Copernican revolution, the initial reference to a non-earthbound point was compatible with an indefinite number of other references, each of which would fling us further away from the earth. But then, under which conditions are we actually at home in our own cosmos? Answering the question concerning the condition of our own home involves, as a prerequesite, that any principle that we may find which will bring together mechanism and finality cannot be used to *explain anything at all* (CJ 412).

Kant also writes: "As applied to one and the same natural thing, we cannot link or reconcile the mechanical and the teleological principle if we regard them as principles for explaining (deducing) one thing from another" (CJ 411). The rejection of explanation in the context of teleological judgment puts a severe constraint on any attempt to compare Kant's predicament with post-Kantian developments in the life sciences. We shall therefore not undertake any comparison of this kind, even though it suggested itself in the case of mathematical physics. Suffice it to say that all formulations of Darwin's theory of natural selection, which aim at disclosing the supposedly tautological character of the concept of fitness, point precisely to the ever-present fuzziness which surrounds the meaning of an explanation in life sciences. Of course, since tautologies are definitions, we cannot forget that tautologies play an important role within scientific theories, making explicit something implicit in the meanings of the theory's key terms. But contemporary discussion focusses on the claim that, when we say that one organism is fitter than another, perhaps no explanation of evolutionary change is embodied at all, but simply a description of it.[44] Let us remember that Darwin's motto was "Natura non facit saltum" (Nature makes no leaps). Even if they are completely independent from the modern concept of evolution, Kant's reflections on the ideal versus the real basis of this principle should still serve as a critical basis for further elaboration.

The single maxim of reflective judgment alone is itself undisputed, but as soon as the bracketing of natural science is effected—and the concomitant need for explanation—we realize how much this maxim, when taken as a firm basis for any further argument, leaves open for disputation. Indeed, the suspension turns the maxim into a problem, "a problem that opens up a wide field for controversy" (CJ 390). In particular, we fall back on the possibility of proof, since we can now ask: "Does the connection in terms of purposes in nature *prove* that nature has a special kind of causality?," that is, a causality different from a merely mechanical one. If natural science had not been put out of play, it would always be possible to reduce the putatively special kind of causality to a highly sophisticated mechanism, as if this mechanism happened to be often too deeply hidden for the type of investigation that we ordinarily carry out in natural science, but never so deeply that mechanical explanation might be acknowledged as absolutely impossible. Whereas the undisputed maxim concerned only our thinking a causality distinct from mechanism (the concept of that causality as a mere idea), the point is now to articulate the problematical character of the idea for its own sake, instead of simply positing this character. The antinomy which follows from that articulation does full justice to the idea as problem, because it makes use of subsumptions under the problematic concept itself.

In such subsumptions, "we do not know whether we are judging about something or nothing, and hence the synthetic predicates of the concept . . . can yield only problematic judgments, whether affirmative or negative, about the object" (CJ 397). Each proposition, which makes up the thesis or the antithesis, *remains* problematical in the following precise sense: the cosmological principle of reason, according to which any sufficiently large sample of the universe must be representative of the whole, is no longer applicable. As Kant puts it: "while [the] maxim of judgment is useful when applied to the whole of nature, it is not indispensable [when we try to discover many further laws of nature], since the whole of nature is not given as organized (in the strictest sense of *organized* as given above)" (CJ 398). However near to or far away from immediate experience we are, an organized being is already a totality. The propositions of the antinomy would be themselves unproblematical if we could construct a totality that would be commensurate with how far experience can reach; this is the kind of construction that reason undertakes in the mathematical antinomies of rational cosmology. But, in fact, each proposition of the antinomy of teleological judgment remains within the confines of the original problem that life is. As we shall see, it either *asserts* the problem (casualism and hylozoism), or it tries to explain the problem by *another problem* (fatalism and theism).

The actual antinomy of teleological judgment results from a modification brought about in the undisputed maxim of reflective judgment which, as we know, sides with the antithesis in the original (but only seeming) conflict between maxims. What is this modification, which has the effect of converting the maxim back into a pair of mutually exclusive determinative judgments? In order to understand it, Kant introduces the notion of *technic*, by which he means nature's procedure whenever "we find something purposelike in nature's products" (CJ 390). The technic of nature is different from the undisputed maxim, in that the primacy of the antithesis in the original conflict between maxims is now used to absorb the thesis as well. By means of the technic, a causality in terms of ideas is attributed to nature by analogy with the principle of art. This causality is an idea of purposiveness which reunites mechanism and teleology, and gives them a new form. Indeed, Kant goes on to divide this causality into two distinct subconcepts, an intentional versus an unintentional one. The former means that final causes must be regarded as a special kind of causality; the latter means that there is basically nothing special about this causality, that is, it is identical with the mechanism of nature.

To the distinction between intentional and unintentional technic, there correponds two interpretations of the rule of purposes in nature's power to produce things; these two interpretations form the basis of all systems that try to account for nature as concerns final causes (CJ 391–95). The unintentional technic corresponds to an *idealistic* interpretation, according to which teleological judgments are not true; whereas the intentional technic corresponds to a *realistic* interpretation, according to which these judgments are true, at least for some of the purposiveness in organized beings (if not for the whole of nature). An analogy with the cosmological antinomy immediately suggests itself, since on the idealistic interpretation *all* purposiveness of nature is unintentional (the unconditioned is the whole series of experience), while on the realistic interpretation one (or possibly several) members of the series play the role of the unconditioned. But the analogy stops here, because each type of unconditioned can be further subdivided into a physical and a hyperphysical version.

Let us immediately note that all of the following doctrines assert problematical judgments, and that this problematical character does not reflect the idea of reason, but the merely sensible evidence of life. Whenever he can, Kant assigns names or historical figures to the various systems under examination. Kant is thus merely descriptive here, unlike in the cosmological antinomy which was entirely part of the critical reconstruction of the world.

Idealistically, natural determination is either casualistic or fatalistic. The casualistic system is physical (Epicurus or Democritus): the causality responsible for the purposive existence of natural things is simply the laws of motion; nature's technic and nature's harmony with our concepts are nothing other than blind chance. On its own admission, the system does not care to explain anything at all, but merely asserts nature's technic, because it does not even wrestle with the question of why nature's products harmonize with our concepts of purpose. The reason why Kant finds it so easy to discard this first system reveals the existence of a remarkable logic commanding the progression from one doctrine to the next; this logic has to do with degrees in accepting the impossibility to explain. For casualism asserts the impossibility of explanation, but defeats itself precisely in that it makes this assertion part of a system. The fatalistic system (associated with Spinoza) is hyperphysical, in that it relies on the concept of an original being. It is much more difficult to discard, because it goes some way toward meeting the requirement of the undisputed maxim of reflective judgment, namely, it refuses to count purposes of nature as products at all. Instead of production, this system admits only connections between accidents that inhere in an original being. What is, on its own admission, problematic in such a system is that the concept of an original being is not intelligible. The connections between the accidents are not related causally to one another, because they do not derive from that being's understanding, but rather from the necessity of its nature. The accidents merely subsist within the unity of a subject, which, therefore, cannot become the unity of a purpose as long as the original being has not caused this connection through its understanding. Reference to an original being is meant to explain purposiveness, but as soon as explanation gets started, fatalism finds itself in the awkward position of having to admit the unintelligibility of the mode in which this being acts. Failure to do so brings it back to natural necessity, that is, the mode of explanation of the physical laws of motion (blind chance). Thus, the problematical character of both the casualistic and the fatalistic systems is enough to reject them. Starting from the enigma of causality, the casualistic system is, according to Kant, obviously absurd, because it puts the enigma out of play (so that the idea as problem is dropped). On the other hand, fatalism maintains the enigma at the cost of dropping any reference to causality: the concept of an original being in this system cannot be made intelligible, for if it were, the connections between accidents would be imputed to the being's understanding, which runs against the system. But if an enigma helps fatalism from collapsing, we are led to suspect that, if there will ever be a way out of the antinomy, this will reside in the fact that there must be something true in the realistic interpretation.

On the realistic interpretation, the irreducibly contingent and indeterminate character of purposiveness is maintained. What characterizes all systems based on this interpretation is that they lay claim to some kind of "insight into at least the possibility of a special kind of causality" (CJ 394), over and above the fact that there is no experience which could ever be adequate to such possibility. Hylozoism is the physical system according to which matter is alive—whether that life, which acts according to intentions, is in the matter or whether it is due to an animating principle. The manner in which hylozoism tries to make its point is paradigmatic of all attempted proofs in this antinomy. Given the problematic character of the proposition, the proposed proof is circular, not apagogic. Indeed, no matter what the principle of organization may be, we could not begin by denying that we have experience of organized beings. Instead, "we try to derive the natural purposiveness in organized beings from the life of matter, while yet we are familiar with this life only in organized beings and hence cannot form a concept of the possibility of this purposiveness unless we have experienced such beings" (CJ 394–95). The circle reduces to a point in the case of casualism, because no explanation beyond what a thing is attaches to that thing; it is a single all-embracing sphere (a substance and all its accidents) in the case of Spinozism. Finally, the last of the four systems is theism. This is the hyperphysical system based on realism, according to which there is an intelligent, living being which, as original basis of the universe, produces the things of nature according to intentions. This being is one member in the series of conditions of the universe, which happens also to be the origin of the whole series. Only theism holds the promise of a way out of the antinomy, because (unlike Spinozism) it ascribes an understanding to the original being which is not eliminated in the act of trying to make it intelligible. In fact, the theistic proposition repeats virtually verbatim the argument of the undisputed maxim, except for the claim about its validity for a determinative judgment. That is, both propositions agree that, in order to account for manifestly final causes, we have to think a causality distinct from mechanism, but the theistic proposition adds that this conclusion must result from the negative outcome of an attempt to discern the unity of a purpose in the material mechanism of nature. We would then have no alternative but to expel outside of matter the principle explaining the unity of a purpose; the principle would be taken in by a supreme understanding as cause of the world. The critical response to theism consists in displacing such a failure from matter to the cognitive powers themselves: because of "the character and limits of our cognitive powers . . . (we have) to give up any attempt to find in matter a principle" (CJ 395) that would send us beyond material nature itself. Theism, in its two versions (regulative and determinative), is thus the point of encounter between two negativities or deficiencies. For the first time in the critical enterprise, a return of subjectivity onto itself is not meant to assert the only possible positive ground of an object (as in the Copernican revolution), but rather to reach beyond itself to a purely spiritual world of higher understanding, regardless of any consideration for an object.

We had already accepted the collapse of the cosmological principle concerning organized beings. The fact is that we simply do not know about organized beings beyond the earth, we cannot assume (except dogmatically) the whole universe to be a vast earth-like system. Now, the foregoing conclusion, drawn from a critical examination of theism, enables Kant to give an a priori basis for this fact. Theism is the last of the four dogmatic doctrines, and thus also the closest to the critique. If, after having so circumscribed the problem of a higher causality, this causality will finally remain a problem, this is because our understanding must be of "special character" (CJ 408).

That the understanding can surpass itself when dealing with natural purposes does not mean that it becomes equivalent to reason. If this were so, the reference to a supreme understanding would immediately reduce to a form of transcendental illusion. Rather, it indicates that, when reason demands the unconditioned and mistakenly converts a reflective maxim into a determinative judgment, its mistake consists in underestimating the action of the understanding. Just as a transcendental illusion results from the unobserved influence of sensibility on understanding, the reference to a supreme understanding, if it is to be acceptable to the limits imposed by critical reason, must result from an as yet unobserved influence of the understanding on reason. What, if any, is the action of the understanding on reason, even though the understanding can never keep pace with reason's aim at the unconditioned? Kant outlines it in the following way: "it restricts the validity of those ideas of reason to just the subject, yet in a universal way, i.e. (as a validity) for all subjects of our species" (CJ 401). Through the search for the unconditioned which is supposed to be the ground of the whole world, our reason elevates us beyond ourselves, that is, beyond the condition whereby we belong to the human species. Nevertheless, it is precisely the understanding's call to order which makes us aware that we belong to the human species. It is a restriction (down to earth) without which any opening (beyond the earth) would lose any basis. Our understanding could not be what it is (in relation to determinative judgment) without being at the same time beyond itself (in relation to reflective judgment). Indeed, for our understanding, to think anything at all means "merely presenting the thing as possible" (CJ 402). That is, the thing may not necessarily exist, which is *more* than what is available to an intuitive understanding, where thinking coincides strictly with being: any thing that is could not possibly be otherwise or not be at all. The intuitive understanding, which conceals the ultimate cause of the world, can be legitimately projected by our human understanding, because "such a being could have no presentation whatever of the possibility that some objects might not exist after all" (CJ 403).

This reference to an intuitive understanding enables the properly human understanding to discover its action on reason. Our understanding proceeds from the universal to the particular, so that the particular subsumed under the universal can never be determined as particular: it contains something irreducibly contingent, because the understanding can only encompass it in a concept which goes beyond its particularity (CJ 406). What makes the particular such or such individual cannot be accounted for when the universal supplied by our understanding is used in order to determine it; as Kant puts it: "our understanding determines nothing regarding the diversity of the particular," because "it must proceed from the *analytically universal* to the particular" (CJ 407). The universal is never more than the formal generality of a concept, and no intuition of a whole as a whole (which would be the *synthetically universal*) is attached to it. In each particular which meets us (through the forms of our sensibility), there is something like a free space, which allows it to cling to some one or another concept of the understanding; it thus lends itself simultaneously to several possible laws of nature. Reason comes into play in order to bring about order in the organization of these concepts themselves: it unifies them according to the supposed law of the unconditioned, and hopes to stop the free play of contingents by connecting each of them to a common origin which can be assigned for once and for all. But in so doing, as we already know, reason is also prone to going too far. A free play, or irreducibly open space, is found again—not, to be sure, at the level of the particulars, since a regressive synthesis constrains them by connecting them to one another—but at the level of the unconditioned itself.

Thus, in the antinomy, the unconditioned always presents itself under two mutually contradictory guises. Reason imposes the lawfulness of the contingent (which is purposiveness), but it fails to prove that our subjective sense of purposiveness overlaps perfectly with any assignable object. Now, Kant argues that, "unless the power of judgment has [its own] universal law under which it can subsume that particular, it cannot recognize any purposiveness in it and hence cannot make any determinative judgment about it." Judgment draws such a law from itself, that is, it discovers the universal appropriate to the particular on the occasion of some "special experiences" through which purposiveness is directly (that is, not mediately, as it would be the case for our discursive understanding) displayed to us in nature.

The free space surrounding each contingent, below the universality imposed by the concept of understanding, was that which allowed reason's cosmological principle to move forward in an indefinite series: even though it is not yet intuited, the subsequent member of the series could be anticipated because it must have features in common with those immediately antecedent which had already been intuited. We now find that this forward movement is itself conditioned by an underlying condition which does not lie in the object aimed at the end of the series. This condition raises the spectre of nonsensicality for the entire process of anticipation. Indeed, anticipation is no longer valid for the object, but only subjectively. "The concept of the purposiveness that nature displays in its products must be one that, while not pertaining to the determination of objects themselves, is nevertheless a subjective principle that reason has for our judgment" (CJ 404). We end up with a new kind of cosmological principle, which is a principle of reason for judgment, not for understanding. Whereas the old principle allowed us to anticipate new members of the series of experience, prior to having been intuited, "our understanding *must wait* until the subsumption of the empirical intuition under the concept provides this determination for the power of judgment" (CJ 407). All reflective judgment can do, in the act of giving a universal law to itself, is surmise that there is a harmony between the particular, as encompassed by our understanding (which leaves it anonymous, as it were), and this particular in nature's diversity. The principle of this harmony "is regulative (not constitutive), but it holds just as necessarily for our *human judgment* as it would if it were an objective principle" (CJ 404). That is, our own subjective power substitutes itself for an objective one. It has the strength to do so, because the new cosmological principle, by forcing reason to apply to judgment, leaves room for the understanding to constitute itself in contrast to "the idea of some possible understanding different from the human one" (CJ 405). In determinative judgment, to represent something as other than it is, or as being absent, requires transcendental imagination; our human understanding, Kant says, is such that "it requires images (it is an *intellectus ectypus*)" (CJ

408). In the case of reflection, the same exercise suggests the representation of another understanding altogether, which Kant names *intellectus archetypus*. This higher understanding is a possible understanding, that is, it may not exist at all. The postulated harmony is thus no more than a peculiar character of our understanding: namely, that it recognizes itself in a higher understanding, for which the universal that human judgment gives freely to itself does in fact correspond to the only possible one for the one world in which we actually live.

In the final analysis, after having acknowledged its weakness regarding the subsumption of particulars under a given universal, judgment discovers its quasi-suprahuman strength in the power it has to give a universal to itself, a universal that is arguably in conformity with the particular in nature. Through the presupposition of a higher, but merely possible, understanding, we discover the special character of our own understanding whereby it can surpass itself. But can this self-surpassing be sustained against the spectre of nonsensicality? Our understanding monopolizes a trait of the higher understanding only on the condition that it renounce another one. The trait that it monopolizes is a purely subjective condition, namely, the coincidence of possibility with necessity: "only by presupposing this idea . . . *must we consider* certain natural products, as to how they are *possible*, as having been produced intentionally and as purposes" (CJ 405).[45] But in order for the understanding to relate to natural products which have been produced intentionally, it must refrain completely from any reference to the nature of the cause that so produced them. That is, we give up implying that there must actually be a special cause that determines products of nature; the ultimate basis that makes such products could still be found in the mechanism of nature. When our understanding says: things could not have been otherwise than they are, it places a restraint on the imagination which allowed it to think itself otherwise than it is. The same imagination, which has enabled us to think a thing as simply possible, has also enabled us to think ourselves as different from what we are; but once we are another, we have given up the power of imagination and its free play of possibles. Our understanding's flight outside of itself has the ultimate effect of bringing it back down to earth. To be sure, it asserts itself in the manner of an intuitive understanding, but it does so only to denounce and abandon, for once and for all, the dogmatic confusion (which arose from an uncontrolled imagination) between the given and the created, between connection and production. Our understanding thinks, but does not create, in the manner of a nonhuman understanding. It can do so without transgressing the critical limits, because this nonhuman understanding occupies the space left *empty* by reason which has already failed to master the absolutely unconditioned (by virtue of the new cosmological principle); it does not result from a demand for an absolute which impels the understanding to transcend itself and *fill* this empty space with pseudo-objects. As a matter of fact, "we do not have to prove that such an *intellectus archetypus* is possible . . . [but] only that the contrast between such an intellect and our discursive understanding . . . does not involve a contradiction" (CJ 408). This proof is the long overdue solution of the antinomy of teleological judgment.

Whoever speaks of proof needs a representation of totality. How does a new relation to totality emerge from the postulation of a higher understanding? Such an intellect would have the intuition of a whole as a whole, that is, it would proceed from the synthetically universal to the particular, whereas our understanding relates to the whole by means of the empty generality of the concept, and our reason by means of the pseudo-filling of objects. The rules that unify nature are no more than subsumptions operating in accordance with the logical principle of parsimony. There is certainly unity, but not real unity, only a unification at infinity by means of this indefinite subsumption, which reaches up toward the most general concept from which all the others can be derived. Formally, our understanding thus proceeds from the whole to the parts, but the most general concept from which it starts could always be another—it is no more than a point (a *focus imaginarius*) concerning which imaginary variations are always allowed. In reality, our understanding always proceeds from the parts to the whole, leaving an irreducible contingency in the combination of the parts, without which different possible forms for the whole would not even be presentable. For us, "the real whole of nature" is no more than "the joint effect of the motive forces of the parts" (CJ 407), an overall effect which can therefore always present itself in other guises. Now, what does the synthetically universal (the intuition of a whole as a whole) available to the higher understanding teach us with regard to the way our own understanding relates to new members of the series of experience? The reflective principle of purposes, in dealing with products of nature, remains a heuristic principle like the original cosmological principle (CJ 411). Even though it cannot know a higher understanding, our understanding cannot but try to *mimick* the higher understanding as the latter proceeds from the whole to the parts. At that level, the power of judgment to legislate itself by means of self-produced principles gives out. What remains is only an *effort* of presentation which never really terminates, because it does not necessarily lead to any fixed determination in any object, not even the object of immediate experience. Domesticating this effort is the zenith of critical philosophy.

Given the peculiarity of our understanding, we cannot have the whole as reality. But we do, nevertheless, have the *presentation* of the whole, which contains the basis that makes possible, not only the form of the whole, but also the connection of the parts. For *our* understanding, knowing means asymptotically approaching the originally infinite forms of sensibility. Likewise, for *our* reflection, judging means asymptotically approaching an intuitive understanding for which the whole of the materially sensible is presented. Hence, the whole, from which the natural products depend as far as their connections are concerned, is only an effect of the peculiarity of our own understanding regarded as cause. But an effect of a cause, which is determined merely on the basis of the presentation of that effect, is nothing other than a purpose. A determination which results merely from a presentation can never be fixed, and hence it cannot be a determination at all, because of the free play of imagination which is always associated with a given presentation. In this case, however, such a play is immaterial inasmuch as we do no more than try to present a whole that makes possible the particular connection—and none other—which is required for all the parts. Have we not, then, spiritualized the whole of nature, transforming the whole into a purpose after the way we present an organized body, namely, a work without respite aimed at resolving a permanently renewed problem? In order for the whole to be presented, we must view it from an outside point through which it becomes a hierarchical system: "as far as matter is organized," we arrive at the concept of natural purpose, which in turn "leads us necessarily to the idea of all of nature as a system in terms of the rule of purposes" (CJ 378–79). What is the room left, then, for mechanical explanation, and thus also for *our* understanding independently of any reference to a higher understanding, an explanation which remains necessary, if only to remind us that we are judging products of nature (CJ 413)?

This cause cannot be an efficient cause, for otherwise we would again think determinatively, that is, we would aim at explaining one thing from another (CJ 411). On giving up determinative explanation, we also give up any kind of mechanical explanation which would be simply a teleological explanation taken backwards. The representation of nature as a whole is a dynamic representation in which the role of time is such that no reversibility of this kind is allowed. Indeed, in surpassing itself, our discursive understanding puts itself in the shoes of an intuitive understanding for which there is no difference between possibility and actuality. Relying on an idea of totality proper to the intuitive understanding (the whole as prior to the parts), it has renounced the use of the power of imagination and its free variations on the given. Thus, a supreme architect of the world could act in either one of two ways: "directly [creating] the forms of nature as they have always been," or, assuming the evolutionary character of the whole of nature, a temporal sequence could do no more than constantly reassert the same, because predetermining the forms that keep developing in the course of nature requires the permanence of "the same model" (CJ 410).[46] In trying to mimick the higher understanding, our understanding has thus forgone the temporal synthesis which characterizes the employment of imagination. This remains perfectly consistent with the critical ground of knowledge. If any time-like condition must be dropped because it plays no role in the global synthesis that we are seeking to characterize, *space* remains the sole means whereby our discursive understanding can think in the manner of the intuitive understanding. "Space . . . resembles the basis we are seeking inasmuch as no part in space can be determined except in relation to the whole (so that in its case too the possibility of the parts is based on the presentation of the whole)" (CJ 409). Now, it is enough to recall that the unity of space will characterize the unity of the basis on which natural formations are possible, if and only if we pay attention to the fact that space is only the formal condition for the reality of products, not the basis responsible for that reality. Therefore, since space can never go beyond the material world as mere appearance, a space-like condition is also inadequate to characterize the inner possibility of a whole as a purpose in terms of an idea of this whole. Having dropped time in order to jump to an intuitive understanding, and having acknowledged that the way space employs a totality prior to the parts cannot but miss any sense of inner possibility of a whole, we are left with the conclusion that, even though an idea of the whole must be presupposed, "it does not follow from this that it is impossible for (an organized body) to be produced mechanically" (CJ 408).

Through the long detour of reflection, we have reached this conclusion, seemingly equivalent to natural science, but without the mediation of natural science. This movement of reversion to the mechanical explanation is thus the counterpart to the movement that justifies the critical transition to the transcendent. Organized nature manifests an organization beyond the categorial organization imposed by the understanding on the raw material of intuition. We began the investigation of this material by trying to adjoin a teleological explication (as index of the organization proper to nature) to the mechanical mode of explication and its categorial organization. We could not separate mechanism from teleology, because the subsumption toward the infinite, whereby the formal unity of acts of the understanding aims asymptotically to rejoin the real unity of nature, only makes sense in terms of the concept de purposiveness. Once the explication of these natural products has given us the idea of such a material purposiveness, we tried to extend the principle of this explication to the whole of nature. In order to do this, it was necessary to introduce, besides nature as a system of laws and of physical objects, but in opposition to the cosmological principle, the notion of nature as a hierarchy of beings and of sensible things, *without considering whether or not its experience has been or will be achieved*. A purely mechanical product of nature is then compatible with a supersensible substrate, which is not an appearance, but which nevertheless could be the thing in itself which conceals the idea (accessible only to an intellectual intuition) responsible for its production. This is exactly what enables us to prove that the understanding's movement, its elevation beyond itself, is simultaneously a falling down again to its earth-like origin.

Freedom, Contingency, and Non-sense

Let us, one more time, compare the cosmological and the teleological antinomies from the standpoint of their respective "solutions." In the former, a confusion between appearances and things in themselves is pointed out as responsible for the conflict. In the latter, we admit, as a premise, that the appearance derives from the thing in itself, but there is confusion between the given and the produced: because it is often foolish, reason demands an explanation where none can be furnished; mere reflection has already taken possession of the unconditioned. The critical solution of the antinomy of teleological judgment, which has gone out beyond the realm of nature, reveals retrospectively why there was an inevitably dogmatic trap at the outset of our investigation of nature. We can only try, and we cannot help but try, to present the sum total of the objects of the world in conformity with a putative archetypal understanding. As a consequence, "we also do not know how far we may get with the mechanical kind of explanation that is possible for us." That is, "we cannot tell what are the limits of our ability for this way of investigating" (CJ 415). We have then, just at this point, renounced the effort to satisfy the critical enterprise, which is a work of fixing limits. Recall the transcendental illusion of rational cosmology, which persists even after its detection; it is precisely due to this persistence that we could assign definite limits to reason's most pressing demand. But it is also precisely because, in the case of natural purposes, limits are no longer assignable that our human understanding could produce, by using its own finite resources in intuition, a nonhuman understanding. Can the latter answer for the finitude of the former?

Let us go back to the fundamental distinction between determinative and reflective judgment. The former is the judgment of experience: the universal is given, the particular must be found. In the latter, only the particular is given, and the universal must be found for it. As for teleological judgment, it is reflective in the following sense: it is "the ability we have to subsume the particular laws, which are given, under more universal laws, even though these are not given" (CJ 202, first introduction). Suppose that the original source of a concept of the understanding is found in the power of judgment, "it would have to be a concept of things of nature insofar as nature conforms to our power of judgment." This concept is nothing other than the concept of the purposiveness of nature. There is something like a Copernican revolution at work between the power of judgment and the concept of understanding: the concept conforms to the power, instead of the other way around, because without the power of judgment, the understanding could not orient itself in nature. But, in contrast with theoretical knowledge, this reversal has the effect of removing any necessity. If there is anything like a concept that expresses a kind of preestablished harmony between the understanding and nature, this concept would not fall within the purview of the understanding, that is, this concept would not prejudge anything as regards the *contents* of empirical laws. Thus, the more universal laws, which are not given, *could be wholly other* than those of purposiveness. Unlike the category, the concept of purposiveness is *not itself necessary*. By virtue of nature's technic (the concept which arises originally from judgment), even if nature were completely contingent, its laws infinitely diverse and its forms infinitely heterogeneous, we could still produce laws of regularity. This shows that the concept of purposiveness is something forced on nature, a position of authority (Kant says that judgment acts "for its own benefit" [CJ 203, first introduction]) which is no longer necessarily at one with our ability to make do. Kant had begun by arguing that a logical necessity existed in given experience, since a universal (provided by the category) is always already there. But beyond experience, and notwithstanding the logical necessity of the cosmological principle, there is no longer any logical necessity, *not even an illusion* of logical necessity that could be corrected in some way: the concept at the basis of the power of judgment could be entirely different from that of purposiveness. In the proofs of the cosmological antinomies of pure reason, the *contents* corresponding to a given rule can never be found; in the antinomy of teleological judgment, the *rule itself* cannot be found, because once found it can always be thrown away by its opposite.

We have followed the intricate pattern of Kant's doctrine of teleological judgment, and we conclude that the absolute fragility of the concept of purposiveness is the only possible way of combining the specter of non-sensicality and the desperate reference, made by our own understanding, to a higher understanding. We will get a measure of Kant's boldness in this respect, if we compare it with a more contemporary argument. In our own time, we have the example of Einstein, who argues that those physical concepts which succeed in bringing about connections between particular laws are free creations of the human mind; they follow criteria such as economy of thought, aesthetic elegance, and the like. But at the same time "this universe of ideas is just as little independent of the nature of our experiences as clothes are of the human body."[47] Likewise, Kant's concept of purposiveness is actually a free creation of the human mind, but it has the peculiarity of prejudging absolutely nothing as regards the content of empirical laws.

And yet, since the power of judgment must presuppose itself as regards the universal that it seeks, it cannot but project itself onto experience. Indeed, in order for "judgment [to] subsume the particular under what is universal though still always empirical" [CJ 203, first introduction] (because the universal is still sought), experience needs the systematic organization of its empirical laws. We run up against a radical impossibility of establishing a first foundation: the founding act, as it were, always chases after itself in reflection. The power of judgment puts itself to work only when it is already at work; it is in its activity that it recognizes itself. Thus, it is not the lawfulness of judgment alone that accounts for the lawfulness of purposiveness. It is as though the freedom that nature chooses for itself corresponded to the purposiveness we put in nature. But this correspondence remains contingent, and so the principle that reflection chooses for itself cannot quite escape the threat of arbitrariness. We would have an antinomical conflict of the type of rational cosmology—and Hegel would thus be right—if the power of judgment necessarily had to assume that the purposiveness that it projects was realized objectively. But the power of judgment *does not need* to suppose nonpurposiveness in order to assert purposiveness. This power is endowed with an absolutely unconditioned freedom to posit anything it likes (this is its heautonomy), so that in the antinomy of teleological judgment it can limit itself by claiming (in the antithesis) that only some products of nature must be explained in accordance with purposiveness. It cannot limit itself by positing something other than itself, because the mechanical explanation (in the thesis) cannot even be denied.

The Problem of the Future Reconsidered

In Kant, reflective judgment makes itself impervious to error by claiming to have in its grasp an unconditioned that it has not found outside itself. In this way, reflection does not infringe upon the purely formal character of superhuman infinity in determination. Only the future of reflection (history and its vicissitudes) will show whether the concept of purposiveness was necessary. But by giving itself the law now, the faculty of reflective judgment dogmatically extends the present to all future eternity. Paradoxically, its assertion of freedom (through heautonomy) cripples freedom for all time to come. Fichte brings reflective judgment back into the sphere of error and truth through the intermediary of what he calls an obscure feeling.[48] Instead of considering freedom (in heautonomy) to be a nascent reflection that is always reiterable—because the corresponding universal will always be missing—we can give the superhuman unconditioned a real weight. This weight so overwhelms reflection that, in the extraordinary profusion of laws, reflection can no longer find itself. Lost in a morass of rules, judgment cannot avoid confusing one rule with another.[49] It must therefore descend into the series of conditions and lose itself in the play of positing and counterpositing. Instead of placing the superhuman in the human as evidence of the unreflective and insatiable desire of reason, Fichte asks us to engage in a deliberate ruse with the infinite to facilitate its entry. We should then be able to find the key to the initial problem of all transcendental philosophy, which consists in the unprepared irruption of the infinite as the framework of all finite experience.

We have to reflect in such a way that that which exceeds us (the not-self) does not exist, though it is still there.[50] This regulated reflection is the work of the imagination, by means of which a repetition (*Wiederholung*) of the self's act of self-positing opens it up to an exterior influence, a check (*Anstoss*) initiated by nothing determinate.[51] If one can distinguish the activity of the self from its product, the repetition of the same can produce something new. One can then receive an *x* which was found in no past synthesis or antithesis. How, then, are we to reconcile, on the one hand, the encounter with this check, and on the other hand, an act of understanding that lies entirely in the self? Fichte undertakes something like a bracketing of the Copernican revolution (in Kant's sense). Since the understanding and the encounter are not identical activities, the question still remains "which of the two is to align itself to the other."[52] The answer is clear. Everything must depend on the absolute self. Since the absolute self posits itself, in the relation to a possible object, it agrees only with the *form* of this object, and not its actual identity. In other words, in its relation to the object, the self discovers itself to be an infinite effort, tending toward mastery over the form. This discovery is, however, a result of the bracketing and not its condition of possibility.

Of course, effort (or tendency) has no causality; it is not primordially determined by any object, but we can reflect on it *as if* it had a cause. Indeed, by virtue of reflection, the self at least exercises some "causality upon itself."[53] Instead of claiming that reflection is not directed towards anything determinate, one might just as well say that in its absolute freedom it does have to direct itself toward something, even if only unconsciously. The direction is provided by conformity to duty.[54] Tendency does not produce an object, but this conformity does explain *feeling* as a determinate tendency, a push toward an unknown, which is nonetheless something.[55] While Kant refuses to oppose nonpurposiveness and purposiveness, Fichte opposes activity and nonactivity in terms of consciousness and unconsciousness. The self does not see the production of the object it produces; this production is unconscious. The self is thus led to deny its own activity temporarily.[56] The activity is restored by means of another act (a new reflection) which has the peculiar property of interrupting the act of determining the *x* that is foreign to the self.[57] This new doubling of the self's positing has the effect of giving the form of the *x*, which is produced (unconsciously) by the self, the character of an *image* of this *x*.

Clearly, there are several images possible for each form, so that the product of the self "is posited as *contingent*, as something that did not necessarily have to be as it is."[58] One moves from the form to the image of the form in the same way one moves from feeling, which is related to the real without seeing it, to intuition, which does see, though ineffectually. How to make feeling see? The self could not even live if it had no feeling of anything. But how can it direct itself toward something determinate if it is an irresistible push toward the unknown? In response to this question, we will have to say that the interruption of the act of determination indicates another tendency in the self, "the drive toward change of feelings," which Fichte defines as the object of *longing*.[59] Though it is no more determinate than feeling was, the change of feeling at least removes the blindness from feeling. Instead of not seeing at all, it can see without being able to account for the act of seeing. Finally, it can legitimately pretend to see, for it does so in accordance with the self's freedom: "the other feeling that is longed for, the self can determine it in itself as it pleases."[60] That which is different from the unknown remains equally unknown, but the self can operate as it pleases between the two with the satisfaction of not violating the rights of the final object it infinitely aims at without ever knowing this object. The arbitrariness of our freedom no longer annihilates the higher freedom of the absolute, nor is it annihilated by it. Feeling is now accompanied by approval. Through the yearning that I experience, I also experience the feeling of being drawn up by a world-projection in which all the objects of tendency converge. This project does not exist, for it has moved ahead of me into the warp and woof of the world. Yet, it is there, since it draws me up and constantly pushes me into a future that is not just anything at all.

The whole deduction of feeling and intuition has taken place as if in a dream. Of course, the dream is a traditional sign of the future. In it, I am possessed by a meaning which escapes me. But even while dreaming, I feel that if the dream were interrupted, I would become the master of meaning once again. When Socrates asked the question of what science is, before concluding that the ultimate elements of all knowledge are themselves unknowable, he answered one dream with another.[61]

In this way, Fichte seems to have overcome what, in the Transcendental Dialectic of the first *Critique*, Kant called the vice of inverted reason. Reason does indeed attempt to determine what the understanding, in its empirical employment, is unable to determine. By means of regressive synthesis, it proceeds toward ever more remote conditions in the past. But to begin from an origin amounts to hypostatizing the reality of the principle of final unity. This would mean that we are dictating ends to nature (A693/B721). Reversed reason can thus be called a vice in the moral sense of the term, for in the moral order it is forbidden to base obligation on a source outside oneself. But Fichte finds a way to reconcile absolute arbitrariness and absolute necessity. What sense does this have for our lives?

In order to discover the first principle, one may begin, Fichte tells us, with an arbitrary fact of consciousness, since consciousness is not yet anything determinate. But what exactly is a primary reflection that is guided by nothing? We must at least have a presentiment of what will come of it. Since we already know that our decision will ultimately be grounded in a higher necessity, we must begin by admitting that the arbitrary point at which we begin is so only superficially. Thus, Fichte in reality begins from one layer of the world "below" the fact, one that stands beneath the immediate appearance of appearance. It is not given, and yet it is something that is there, which guides the first contact with the surface. This something that is there is existence, life, though not as an idea, but rather in its very flesh.[62] The absolute self is like the body. Inside there is a given that exists, but which is not really there for me. I experience it most often through pain, or just the absence of satisfaction. On the surface, however, there is the skin; something that is there, but which one can remove, turn over or distort almost completely arbitrarily. Let us call this "there," which is never given once and for all, the *texture* of my world. One can then understand why, correlatively, in the *Science of Knowledge* the ultimate horizon of reflection is yearning. The self first operates as it wishes; it can posit an absolute which, as a world-texture, lends itself to all manner of manipulation. Yet the arbitrariness of the movements on the surface of the body disappears, and leaves room to the inescapable, once one penetrates more deeply into it. Similarly, the self acts from the outset (though without knowing it, i.e., by means of imagination alone) in conformity with the real givens of the world, which exceed its private sphere: the ultimate (unknown and unknowable) ends of rational freedom. When the self opens up to the Other, it only discovers what is always already given in itself without being there. It is therefore the absolute totality, which contains at once everything in it and outside it.

Is there anything comparable in Kant to the texture of my world, something which would afford critique the opportunity to save face before the danger of non-sense that undermines it? Must we really engage in a Fichtean-style radicalization of critique? There would seem to be something equivalent to the texture of my world in Kant's doctrine of right. The doctrine of right deals with external freedom, that is, the conflictual contact of *several* bodies and the work of history that arises from this conflict. The relationship between the given and the texture is the relationship between internal and external freedom, mediated by the doctrine of right. Before we can broach the Kantian sense of texture, we must return one last time to internal freedom, for it is what first brings us out of a reflection that is caught up in the cosmos.

CHAPTER 7

Lost Illusions

Renversez ces propositions? Vous arriverez au vrai.

—*Balzac*

On Concepts Other than Cosmic

Given the absolute fragility of the concept of purposiveness, is the critical enterprise doomed to fail? Not if we can discern, in a critical manner, a class of concepts *other than the cosmic*. We must return to this threat of collapse at the point where we left it, this time without leaving the Kantian compass. After the Dialectic of Teleological Judgment, Kant goes on to distinguish our "*authority to try*" (which he finds to be absolutely unlimited as long as it stays within the confines of the mechanical terms) from our "*ability to make do*" (which is limited when we restrict ourselves to these confines) (CJ 417). In order that our authority be one with our ability, we would have to think of the same mechanical terms as both limited and unlimited. What allows such conjunction is nothing other than the teleological mode of explanation, which is consistent with the fact that the mechanical route never leads the natural scientist to an impassable wall. It asserts that the point at which the mechanical route gives out does not arise from an intrinsic impossibility originating from within this route.

Our understanding is offered the possibility to present a whole after the model of a higher understanding. But then, if nature is nothing other than the object of the idea, it is a short step to arguing that the dynamical character of nature can be entirely controlled by the idea. This interpretation, which forms the basis of German Idealism, makes light of the necessary reversion to mechanical explanation that Kant articulated at the end of the Dialectic of Teleological Judgment. Nature would then be objectivized reason, or reason in the process of objectifying itself. Kant always speaks of the object *in* the idea, that is, the ideal object which never lends itself to any determination, as distinct from the object given absolutely (A670/B698). Is there a dynamical character of nature that refrains from twisting the object in the idea into an object given absolutely? The answer is yes: there is a reality, other than the totality, which can be represented within critical limits. This reality is the human being as noumenon, the source of freedom and moral ends. The third cosmological antinomy of pure reason is the place where this other reality is thematized for the first time. A specifically human freedom can no longer be dependent upon an intuitive understanding which does not know of the rules employed by our own understanding. This is transcendental freedom, which can assert itself only through a conflict with nonfreedom.

Kant does not, however, directly oppose such a transcendental freedom to the heautonomy of the power of judgment. He opposes a practical, not a transcendental, freedom to purposiveness. Practical freedom and purposiveness indicate two movements which are the inverse of one another. Purposiveness results from a deficiency in the hyperphysical, since for an intuitive understanding there would be no difference between the possible and the actual. When the difference between a concept (which deals with possibility) and an intuition (which gives something actual) is abolished, both concept and intuition would in fact reduce to nothing (CJ 402). If there is freedom in heautonomy, this is freedom of and for nothing at all. By contrast, as regards practical freedom, there is a deficiency on the level of the physical (the sensible), for it is possible that a morally necessary action may not take place. Again, only in a purely intelligible world would there be a difference between obligation and action (CJ 403–4). Nevertheless, in this case, annulling the difference between possibility and actuality cannot entail abolishing each of the terms of the difference, for otherwise the very possibility of presenting moral laws as commands *for us* would vanish, together with absolute necessity as an "ought to be." For us, the freedom contained in this intelligible world is thus an excessive (transcendent) concept, the one excess that we must live without foundering in the ramblings of an often foolish reason. In order to think the conditions that bring this excess back to a sensible expression, we need a freedom other than practical freedom—transcendental freedom.

Transcendental freedom is asserted in the thesis of the third cosmological antinomy. While the antithesis denounces the illusion of a freedom that would be "a point of rest" offered "to the enquiring understanding in the chain of causes," the thesis replies that rest consists in entrusting oneself to the laws of nature alone, where "there will always be only a relative [*subalternen*] beginning and never a first beginning" (A445–47/B473–75). But a series of subalternate beginnings, however extensive it may be, never covers the whole of the series of causes that arise from one another. In other words, the thesis asserts the existence of a freedom situated on the level of the totality of nature (where every cause is itself caused), and not on the level of the cosmic concept (or absolute totality determining causal relation). The free event (that which results from a resolve and the consequent action) follows, but does not derive from, a purely natural event: there is a force that acts as though it could escape the unfolding of natural causes. Even the antithesis admits that an infinite derivation, without a first member that engenders all the consequences, "cannot indeed, in respect of its possibility, be rendered comprehensible" (A449/B477). But in order to prove determinism, it relies on the analogy with the situation in natural science, where the fundamental forces (attraction and repulsion) must be posited, but remain unintelligible. Transcendental freedom thus occupies a place left vacant by the forces of nature themselves—by the unintelligibility of certain fundamental forces. This reminds us of the way reflection on an intuitive understanding occupies the place left empty by reason, inviting human understanding to push its conquests beyond the domain of its possible experience. Transcendental freedom as it were drifts between the two conceivable totalities (pure nature and absolute totality), without ever fixing itself in one of them, freedom being what ensures the passage from one totality to the other. The spontaneity of transcendental freedom thus consists in the power to bring about actions without being influenced or determined by grounds or motives not produced by this freedom itself. But there is no guarantee that such particular action imputable to transcendental freedom (instead of the causal sequence of nature) is the uniquely determined outcome of our purely rational volition. Because the space between the two totalities remains undeterminable, we can never distinguish which events were produced by transcendental freedom and which not. There is no way to decide which natural events proceeded from transcendental freedom and which are wholly accounted for by the causal sequence of nature. So long as we do not know what sort of being this space is, what its ontological status is, we cannot give an account of spontaneity such that actions issuing from the exercise of the transcendental power are recognizable as such. Free causality thus puts a check on the transcendental mechanism of self-recognition, which guarantees the possible objectivity of judgment. But it is precisely this empty space between the thesis and the antithesis which allows us to avoid Fichte's conclusion according to which, if none of the propositions can directly refute its opposite, this is because their conflict derives from an actually recognizable "first principle," even though this principle "admits of no derivation from anything beyond it."[1]

We now see why practical freedom rests on transcendental freedom. Practical freedom expresses the above-mentioned original indeterminateness by making it sensible, but it does not make it recognizable. The principle of practical reason is that "although something has not happened, it *ought* to have happened" (A534/B562). On the phenomenal level, there was a *lack of determination* in the cause, a lack which the antithesis admits (i.e., by itself the infinite series without a first member is not intelligible). But the antithesis immediately covers over this lack by taking the sum of appearances to be something valid in itself. In fact, the antithesis has not yet separated itself from the common, pre-critical, presupposition that takes appearances to be absolute reality, and falsely makes one think that "nature will then be the complete and sufficient determining cause of every event" (A536/A564). The thesis and the antithesis agree at least that what unites them—provided that one takes care to not confuse the appearances with the things in themselves—is the unsurmountable enigma of free causality. Practical freedom is indeed the unfathomable thing the *Critique of Practical Reason* spoke of: as an unexplainable fact, it can only be the object of pure reflection. The confusion of the appearances with the things in themselves is very different from the fact that "appearances . . . must themselves have grounds which are not appearances" (A537/B565). Where the confusion has the effect of brushing away the enigma by making it harmless, the living reference to the enigma, which is precisely what a free act is, bestows a real strength upon the unfathomable. Humans recognize themselves as authentically free in this unfathomable element. That a human is free in a world which is noumenally valid *only for him* (and not for another understanding, as in the case of heautonomy), means that he lives from the enigma of his freedom, and finds in it the strength to tear himself away from the hold of nature, from the determinism that fixes once and for all—and which is in the end nothing else than a prefiguration of death.

Fichte criticizes Kant on just this point. Kant's unfathomable is only a prefiguration of another world, a transcendent world completely separated from the world we know. Thus, the categorical imperative would be one such determination of the unfathomable. According to Fichte, the transcendent world is both useless and false. He takes the inexplicable fact of freedom to be always *identical to itself* in consciousness. Consequently, as he puts it in the *Sittenlehre*, thanks to the intelligibility that the categorical imperative now receives, "the air that it had, and which is fitting for a mysterious quality (*qualitas occulta*)—this naturally without the positive consent of the author of the Critique of Reason—vanishes completely; and the obscure region, fit to serve as a springboard for every form of enthusiasm . . . is completely destroyed."[2] Once more, by undercutting enthusiasm at its root in the ultimate mystery of freedom, Fichte makes himself the champion of Kantian doctrine. Has he not, however, exceeded the rights granted by the critique? Is identity an authentic intelligibility?

The Texture of Our World: Oscillating between Givenness and Nothingness

The idea of world, fraught with antinomy, is fragile and open to inconsistency. Whereas the mathematical cosmological ideas as it were dissolve of themselves (each pair of propositions is false), the dynamic cosmological ideas call upon the force of an individual to tear himself out of the natural conditions in which he lives. Ultimately, the displacement of the idea of the world to the human world of *culture*, that is, the elevation of the cosmic concept of philosophy to its cosmopolitical dimension, opens the prospect of a greater solidity for it. The human world of culture is the theater of political action. To be sure, political action is dependent on freedom, and thus on the unfathomable; in addition, the world of culture is still the encounter of several freedoms in their sensible dimension. But in the name of the unfathomable one does not do just anything—that would make of history simply a movement in which forces either exhaust themselves or renew themselves at will, depending on the circumstances. (By contrast, the concept of purposiveness *could* certainly be replaced by anything else, since my life incarnated in a body *is* an exhaustion of forces.) To the contrary, the maintenance of the world of culture requires that forces be released as the tasks require them. The human being as a citizen of the world (the cosmopolitical concept), understood as an intersection of diverse forces, must have the will to situate himself in what surpasses nature. His freedom makes of him an end of absolute worth, an "ultimate purpose [*Endzweck*]" (CJ 431). To project this end into nature is the final end, and it organizes a world of culture definitively separated from nature.

Kant thought that the transition from nature to culture, and so the eradication of all shining misery in modern times, would be achieved through the aim of a cosmopolitical whole. A global culture would account for the emergence of morality in humans, and thereby resolve its antinomical elements at the origin (what Kant called the unsocial sociability of men).[3] Indeed, it is only in relation to the species, not the individual, that the incomplete development of a human being's natural dispositions as an individual—which constantly threatens to reduce his limited horizons of certainty to absolute negation—could be transcended at last.[4] But there is nothing naïvely optimistic in this account. Kant believes that the conversion of society into a moral whole, if it is ever to occur, will be "pathologically extorted."[5] Furthermore, as we already know, there is no privileged viewpoint from which this development could be judged: that is, there is no Copernican revolution in history. The conflict in global nature is transcended by global culture at the cost of losing the individual subject.

But there cannot be world citizenship without an organization on the level of law. Whether it be in matters of private law, public law, or cosmopolitical law, conflicts—between individuals, in the relationships that constitute a social group, or between states—have to be regulated. The cosmopolitical level is the level of relations between states. It is, however, only *analogy* that enables us to pass from public law to cosmopolitical law: individuals are violent among themselves and this violence calls for the state of law; states are also violent among themselves, and require a cosmopolitical society. The analogy here indicates not only the limits of reasoning; it also indicates concessions required by individual freedom which has hitherto been represented as a will of separation. The individuals are regulated by what surpasses them: the state. The universal state that would be set up to regulate these conflicts would, however, because of its universality, have no real force, since it would always come up against the will of individuals; it would itself reinstigate adverse factions, and would thus render every concrete project for perpetual peace unrealizable.[6] In his project for perpetual peace, where he advocates a universal form of republicanism in which all states would be seeking peace, Kant gives only empirical and technical arguments against the possibility of a state of all states. In particular, the state that would cover the entire earth would be inefficient in virtue of its sheer magnitude. What differentiates the cosmological from the cosmopolitical idea is thus that the latter is impossible, not because it is contradictory, but because its unrealizability can only be thought in terms of empirical images—it cannot be demonstrated a priori. On this level, we would have to declare that the force of freedom will be defeated. Every cosmopolitics advances toward the unrealizable, not because at a certain moment of history the continually increasing immensity of the task would force humans to renounce it, but because at every moment of history and every point of the earth the humans can never sustain, in the reality of their lives, the weight of the common enigma that defines their horizon and their ideal—the fathomless character of freedom. Maimon announces the failure of Kant's cosmopolitical project when he writes that "our thinking being, *whatever it may be*, experiences itself as a citizen of an intelligible world."[7] Whatever it may be: the preliminary condition of communicability has been wiped out, the unfathomability of freedom has lent itself to complete negativity. Maimon has jumped in with both feet into the dream world of the pure play of thought that Kant denounced entirely as a loss of world.

The *ideas*, however, offer one last chance. Unlike the outward movement of the spectator (the Copernican revolution), in which subject and object stand against each other, the idea of reason is a purely inward movement, an inward growth (like that of a living being) which makes possible the progress of knowledge, that is, the passage from the simple aggregate of different pieces of knowledge to their systematic organization (A833/B861). Kant calls the effectuation of this movement the "supreme end" of reason (A851/B879). A last possibility of peace thus presents itself, once again on the purely theoretical level. We have seen how the hypothetical employment of reason gave rise to a transcendental principle of reason, according to which the systematic unity in the rules of knowledge could be assumed to be necessarily inherent in the objects of knowledge as well. Objective and hypostatized ideas can then be *admitted* (A673–74/B701–2), if at least care is taken not to ascribe any thing to be determined by means of them. That is, their reality can be admitted as schema of a regulative principle. But we see here the antinomies being evacuated or evicted, when Kant tells us that in fact only the psychological and theological idea are objective and hypostatized ideas. The cosmological idea, which remains subject to an antinomy, cannot be an objective and hypostatized idea. As a matter of fact, we have to distinguish between two kinds of employments of hypothesis in the realm of speculative reason, the dogmatic versus the polemical (A769ff/B797ff). The dogmatic employment is ruled out on two grounds. On the one hand, it is sheer laziness, because a transcendent principle of explanation, being preconceived, would serve only as cheap satisfaction for reason; it would fail to further the employment of the understanding in respect of objects. (In this respect, the cosmological principle must be said to be regulative in the sense of "heuristic fiction" as at A771/B799, not hypothesis.) On the other hand, an hypothesis accounting a priori for its given consequences cannot easily get rid of a wealth of auxiliary hypotheses (mere fictions), precisely because the totality of given consequences can hardly be determined a priori. Therefore, only the polemical employment of hypotheses is permitted from the critical standpoint. This means that they are used essentially "for the purposes of defending propositions" (A776/B804). Polemics is the art of using arms one does not necessarily believe in, putting forth opinions one abandons as soon as the defense has been assured. For the needs of defense, we can outright *invent* arms (A780/B808), or even use *dreams* as a polemical arm, even though the dreams elaborated by the imagination are nothing but so many indices of indolence (A770/B798).

We see here how Kant is led to play so dangerously with the arbitrariness of what is unknown and unknowable. This polemics is presented after the critical solution of the antinomies, which is a brand of polemics which still takes itself seriously. We then return to war, as though the critical solution had only transformed a real war into the parody of a dispute. For, to be sure, endless fights result from this employment of hypotheses, but this is now a war using "leaden weapons" (A778/B806). Behind these combats that are not real combats, and the promise of a peace that is always delayed, again something very serious lies to be uncovered. Indeed, "though a merely negative attitude to the assertions of others may seem very modest and moderate, to proceed to represent the objections to an assertion as proofs of the counter-assertion is to make claims no less presumptuous and visionary than if the positive position and its affirmations had been adopted" (A781/B809). We recall that to show oneself better than one is is, in every age and place, an essential factor of civilization, because the mere pretense at the better is the first way for a man to become finally an authentically moral being (A748/B776). One can then say that polemics is the means to reenact indefinitely the (indeterminable) act by which human beings have left the barbarous state of nature; the harmless and apparently groundless games in which polemic is engaged recall the fragility of all human pretensions (including those of knowledge), and thus of human nature and condition in general. The seriousness of polemics reminds us that the texture of our world is not so superficial as it seems.

Transition to the Absolute

The only point in critical philosophy where the excess of sensible experience over any possible determination is actually mastered by means of thought is a special kind of idea, the *aesthetic idea*. In fact, an aesthetic experience (of an object that we deem beautiful) is more wholly subjective than any other experience, because what sensibility gives in such an experience is not its own form (space and time), but the form of the object of intuition itself. As Kant puts it: "[P]leasure is connected with mere apprehension of the form of an object of intuition" (CJ 189). The fundamental fact of cognition, namely, that thought cannot but go outside of itself in order to determine something—even if nothing is actually given in an experience—receives a sensible expression, because this outside is now the medium whereby *communication between subjectivities* takes place. If I find something beautiful, I demand that others agree with my judgment. This "universal voice" (CJ 216) is so pressing that I will even blame others for not agreeing with me, so much so that I will deny them taste if they fail to do so (CJ 213). The demand goes immediately beyond experience; I do not wait for others to confirm my judgment before ascribing it to them. An aesthetic idea is thus the counterpart, the negative figure, of the rational idea: whereas a concept to which no intuition is adequate defines the latter, the former is an intuition for which no adequate concept can be found (CJ 314). This does not mean, however, that thought dissolves and leaves room to irrational demands. On the contrary, the aesthetic idea means a presentation of the imagination, so that *more thought* is contained in it than could be arrived at by means of the rules of cognition alone. The presentation overflows the categories, so that the categories are now themselves expanded beyond limit, and the power of reason is set in motion. Ideas of reason are now given a "semblance of objective reality," because the presentations of imagination, not reason's own hopeless demands, have pushed reason beyond itself.

What does the expanded concept give one to see? It does not simply exhibit itself. It brings about, and expresses, *relations* between categories that the categories themselves (in particular, the category of relation which gives rise to the three ideas of reason) are unable to think. That is, the aesthetic idea expresses "the concept's implications and its kinship with other concepts" (CJ 315); the other concepts become the predicates of the object deemed beautiful. Finally, the effect of an intuition on the understanding appears in its full significance, when this intuition is isolated from the understanding. Such an intuition, which can be aesthetic only, gives the understanding more thought than it can have when imagination remains subjected to the understanding. The understanding expands without limit, even though the understanding, when considered originally in independence from sensible intuition, was acknowledged to be already infinite. The intuition is not even taken in by reason, but rather by the faculty of judgment. Therefore, whereas outside the context of aesthetic judgment the schematic activity of the idea of reason came to nothing, because no object could be determined, what the aesthetic context provides *is* the sought-for schema. But since there is more to think than could ever be determined, judgment treats the intuition "in a way merely analogous to the procedure it follows in schematizing, i.e., the treatment agrees with this procedure merely in the rule followed rather than in terms of the intuition itself" (CJ 351). The rule which does not anticipate anything (the cosmological principle) vanishes for the benefit of an analogy with a rule; the analogy discloses the character of the rule, it allows the rule to see beyond any given content; the procedure of the rule being so exposed, the rule anticipates more than it would otherwise.

Does that surplus give reflection the guarantee that the principle it gives freely to itself is not quite so arbitrary after all? In giving the rule to itself, reflection cannot but miss the reference to the object of intuition that would validate the rule directly. All it can do is thus reflect on that very intuition that it cannot have, which propels it "to an entirely different object, of which the former is only the symbol" (CJ 352). The object beyond is related to the content of the given intuition as the character of the rule is related to the rule itself. Thus, this supplement to the original intuition is not a new object, but a *symbol* for that object. Because the outside of any possible experience is now related to the inside by means of the same rule (concept), the transcendental appearance can no longer be located anywhere in space and time. This does not mean that the necessary illusion is definitively done away with; rather, it cannot even be detected. But at the very least, our ability to reflect on a missing intuition proves that not every substitution (i.e., not every symbol) will do to satisfy the aesthetic pleasure involved in any reflection.

Symbolism thus provides an enlargement of the scope of schematism, because the symbol seems to supply the image that the transcendental schema had to transcend in order to sensibilize such concepts as pure concepts of understanding. Kant calls hypotyposis that general process (*Versinnlichung*) whereby a concept is made present in sensibility; it is either schematic or symbolic (CJ 351). Thus, the hypotyposis is not a proof of the reality of our concepts, because this reality is already secured by the fact that the understanding (in the case of schema) or reason (in the case of symbol) has formed the relevant concepts. Rather, it is a *test* of that reality thanks to which we exhibit whether or not such or such symbol is adequate to such or such reflection, when the reality of the concept is not yet visible (recognizable) in the corresponding universal. Suppose reflection *succeeds* when it gives the rule to itself: in this case, the not-directly-intuitable object and the symbol are held together by means of the aesthetically pleasing similarity between the rules by which we reflect on the two. This is the point at which the whole experimentation with concepts, which began with the exhibition of the transcendental ideality of space and time in any possible cognition, is now brought to an even more primordial level, that is, the level where the universal contained in our concepts is not yet fixed to any determinate law.

The symbol sends us from the impossible form (that we seek in reflection) to a possible form by proxy. The substitution passes over what would be the absolutely formless, the dissolution of every recognizable form, that would otherwise result from the original impossibility. Now, of all the aesthetic feelings that we have, it is the feeling of *sublime* that comes closest to characterizing the decisive moment prior to the fixation of the concept. Recall that the logic of illusion in the Transcendental Dialectic of the *Critique of Pure Reason* is such that, in order to be tested, the transcendental ideality of space and time had to be temporarily suspended in order for it to be finally admitted as impossible to avoid. Similarly, the feeling of sublime captures the dissolution of every possible form (magnitude) from the standpoint of the Ideas of Reason, just prior to their being admitted as reality for the practical law. Thus, the sublime refers to an object which, in virtue of its absolute superiority over us, prepares the mind to accept the humiliation resulting from the failure to attain the totality of nature (CJ 268). This is the moment when the mere claim to everyone's approval in the aesthetic feeling is about to be converted in an object of moral feeling, that is, an outright command for everyone to approve (CJ 267). In morality (*Sittlichkeit*), reason must do violence to the inclinations of sensibility (CJ 269); similarly, in the feeling of sublime, the fearful vision of the formless and the chaotic cannot be sustained unless imagination does violence to our own inner sense (CJ 259). But in so exerting its dominance over inner sense, imagination determines itself purposively, and so it also deprives itself of the freedom it has with respect to the empirical employment of the understanding. In this way, imagination has responded to a call (*aufruft*) whereby it transcends its attachment to nature (CJ 262); it finds within itself a strength which does not belong to nature at all, it puts up a resistance *in thought* against the fearful (CJ 260), and thus develops what Kant calls the supersensible vocation of the mind. In the experience of the sublime, the mind behaves as if it were an intuitive understanding, because its free imagination has been relinquished. What happens to the imagination when it reaches such heights?

Consider the mathematical estimation of magnitudes in accordance with numerical concepts. In any determination of how large something is by means of numbers, there is still freedom as far as the determination of the unity is concerned (CJ 251). In order for the unity to be fixed as well, we need a first or basic measure (*Grundmass*) for which no determinate concept of a given magnitude will do. This first measure must be allowed of being taken in directly in one intuition, an intuition that will serve for the exhibition of all subsequent numerical concepts. The mathematical generation of magnitude that never knows of any maximum is done progressively, whereas the generation of the first measure is done comprehensively (CJ 254). Now, Kant argues that the estimation of magnitude which is not determined objectively but only intuitively is *aesthetic*, because it comes up against a *maximum*: when this maximum "is judged as the absolute measure beyond which no larger is subjectively possible . . . , then it carries with it the idea of the sublime" (CJ 251). In this emotion, that no mathematical estimation of magnitude could ever convey, the mind enters in competition with, and attempts to measure itself up to, the original infinity of its own pure a priori forms of sensibility. Indeed, not even the infinite space and past time is exempted from the demand of reason to comprehend in *one* intuition *all* the members of any given numerical series (CJ 254). This demand for complete exhibition is the mark of our *common* reason, which gives rise to the natural antithetic involved in the self-contradiction of an absolute totality without end. If we were to limit ourselves to cognition, we woud have to transform this demand into a transcendental antithetic—something that has been done in the cosmological antinomies of pure reason. But the *mere thought* of something absolutely large beyond any comparison, without a unity in determinate relation to the infinite, transcends the self-contradiction, even though, of course, it cannot eliminate it. If an intuition corresponded to this supersensible power of thinking, it would have be itself supersensible. Therefore, in abiding by the sublime, our imagination refrains from converting a natural antithetic into a transcendental antithetic, and by the same token it relinquishes its freedom with respect to any empirical employment of the cognitive faculties, in favor of an infinity that goes beyond any standard of sensibility—that is, beyond the very infinity of the forms of intuition in cognition. Because this highest order of infinity, that is, this infinity which can be only comprehended aesthetically and not apprehended mathematically, the claim to universal approval contained in it must also be more than just an invitation to approval; it must be somehow universally binding, *almost* like a moral imperative. The only way to achieve this unchangeability from within the changeability and revisability of our feelings, without using any pregiven law of reason, is to identify the basic measure of nature with the whole of nature itself. That is why Kant calls it "the absolute whole of nature" (*das absolute Ganze*), which is also "*zusammengefaßte Unendlichkeit*" (CJ

255). In so transcending itself from within itself, while passing over the moment of self-contradiction in the antinomies, the imagination has inflicted a violence to its natural vocation in the empirical employment of the cognitive faculties: the self-inflicted displeasure and pain has revealed the true vocation of the mind, which is supersensible. Our minds are traversed from the outset by a supreme tension (*Anstrengung*), which defines reason's natural desire for metaphysics and the corresponding cosmological principle of regression. But the sublime acknowledgment of the inadequacy and powerlessness of our cognitive faculties to ever reach the end of this regression gives rise, at the same time, to the feeling that striving for such an end is adequate to the idea we have of this end. The whole movement of striving to the unconditioned was not arbitrary, but exerted itself in conformity with a law, the realization (*Bewirkung*) of which is the feeling of respect for it.

Prolegomena to Finitude and Death

The absolute of the aesthetic is necessary in order to understand the Kantian philosophy of the human relation to finitude and death. In his Kant book of 1929, Heidegger shows that Kant's concept of finitude must be understood in terms of a summons to a predonation of beings. But this, according to Heidegger, still falls short of what is radically finite in human beings. Kant supposedly determined this concept of finitude in a manner which is still *external*, in opposition to productive intuition—*intuitus originarius*.[8] This interpretation will necessarily miss the thoroughly internal determination of finitude that Kant gives in the *Critique of Judgment*: it is the finite which expels the infinite, in keeping with its own resources, by analogy with, and not in opposition to, the nonhuman understanding. But does Heidegger reach a different conclusion? What is his version of authentic finitude? Internal finitude, which bears upon Dasein in its difference from the animal, which also is dependent with respect to the already-there, is the need of the understanding of Being which manifests itself as Dasein.[9] This transcendental need secures for Dasein its possibility to be, while also allowing us to understand that "more original than man is the finitude of the Dasein in him."[10] Thus, finitude is not an accidental property of human reason, nor the mere lack of creative intuition, but a summons to the understanding of Being, which implies that "there is and must be something like Being where finitude has come to exist." Finitude in itself enables us to account for what is "creative" in the capacities of the finite human creature. Heidegger calls fate the association of the two senses of finitude: as project of being and as dependence upon pregiven being.[11] Dasein in its existence is finite and mortal: it is a being that faces death, and for which its finitude is the very essence of its existence. In the triple structure of time (past, present, future), the future has primacy: it is, in fact, from out of the future, which is to say, from out of the anticipation of death, that the proper (internal) meaning of the temporality of a finite being stands revealed. Heidegger thus comes close to the thesis of the finitude of original time itself, which is opposed to that of the infinity of time. With its freedom for death, Dasein has attained the pure and simple end whence choice concerning factual possibilities becomes possible; and it is this choice which renders it free in relation to the contingency of circumstances, and which transports it into the simplicity of its destiny. For Dasein deprived of the illusions of the "they" before the tribunal of death is more radically attached to its actual situation than inauthentic and irresolute Dasein.

Has not Kant, insofar as he affirms the infinity of space and time as forms of sensibility, simply inherited the classical conception? And does he not irremediably entrench himself in the inauthentic understanding of existence in that he defines the need of reason as that of a never achieved coincidence with this infinity? Heidegger sees the Kantian relation between the finite and the infinite to follow the model of creation, not of analogy or imitation. In fact, the authentically exterior manner of representing to oneself the finitude of the human being, according to the critique, is found in the doctrine of right. Far from lapsing back into the inauthentic mode of the "they," and even though it treats of exteriority, the Kantian philosophy of the relation to finitude and death is found in this doctrine of right. So as to achieve a grasp of this ultimate articulation, one needs to take notice of the fact that the aesthetic absolute is the prerequisite for understanding it.

Feeling and Coercion

Kant's concept of right essentially consists in the law of external freedom, which defines a symmetrical order of coexistence for human beings living according to the spatial relations proper to each society. The law not only delimits the domain which one may reasonably consider one's own and occupy as one sees fit, but it also authorizes one to defend one's domain against encroachments and attacks from without. Indeed, according to Kant, right is *analytically* linked with the authorization of coercion: "there is connected with Right by the principle of contradiction an authorization to coerce someone who infringes upon it."[12] The authorization of constraint is thus the expression of one's permission to defend domains of freedom which are universally compatible with one another.[13] But we may nonetheless ask why Kant does not go on to deduce right, why right, though its concept is not empirical, does not in its universal principle admit of a transcendental demonstration, which would make it a necessary representation.[14] The answer may be found in an analogy; it seems that an absolute distinction between the principle of right and the categorical imperative cannot be drawn, since both contain criteria of universalization that reason draws from itself. In both cases, the other is only postulated; I cannot represent the other's freedom to myself. Each of the deductions of juridical concepts (possession, acquisition, occupation) is an analytic proof, while the juridical propositions to which these concepts contribute are synthetic because, a priori, "practical reason *extends* itself without intuitions."[15] In this sense, the ensemble of these deductions is related to the initial impossibility of a transcendental deduction of right in its universal principle. But if the highest principle of practical reason is indeed a synthetic, a priori proposition, though one that can be only exhibited and not deduced,[16] does the impossibility of a deduction in the case of the universal principle of right indicate that right is fundamentally free of any synthetic character?

We are led to ask whether instead there might not be a synthetic element in right which, for reasons that remain to be seen, stands in the way of transcendental deduction. In order to establish the concept of coercion, Kant relies upon an analogy with mathematics and physics, just as he did in the case of objective knowledge. He writes: "The law of a reciprocal coercion . . . is, as it were, the *construction* of that concept; that is, the presentation of it in pure intuition a priori, by analogy with presenting the possibility of bodies moving freely under the law of the *equality of action and reaction*."[17] It is perhaps the reference made to *space* instead of time which seems at first to justify the rejection of any a priori deduction. For example, one empirical condition to which Kant attributes an unavoidable character, a condition without which the possibility of a community of human beings would be unthinkable, is nothing else than the spherical shape of the earth; on this empirical spatial condition is based the nonempirical, juridical concept of original communal possession.[18] More generally, Kant considers the state of nature that precedes the constitution of a common, public will (the social contract) as a mere idea, and not as a fact which one might assign a chronological origin in the time of history.[19] Kant thus finds that though the concept of right itself cannot be presented in an intuition, this is not the case for the concept of coercion, which, however, we have just seen to be analytically linked to the concept of right. In other words, though right in its universal principle surely cannot be deduced transcendentally, there is no reason for the same to be true of coercion.

While the works of Galileo and Newton were the dominant references in the analogical work leading to the transcendental constitution of knowledge, it does not seem an exaggeration to say that, in the case of right and politics, it is Hobbes who occupies this position. Is it not Hobbes who, in the modern era, hopes to apply the mathematical method to political philosophy in order to transform it into rational knowledge? The problem Hobbes sought to resolve is the following: while pure mathematics never really stands in opposition to the prescientific lifeworld, since counting and calculating belong to both, an exact representation of the political relations between individuals, however, supposes the suppression of the everyday passions which constantly undermine them.[20] For Hobbes, of course, it is a question of conceiving the passage from the state of nature to that of civil society in such a way that passions and natural opinions (essentially focused on the constant play of praise and blame) both lose their sway. In his doctrine of right, Kant presents himself as the champion of Hobbes' doctrine: "The doctrine of Right wants to be sure that *what belongs* to each has been determined (with mathematical exactitude)."[21] Here, the term "exactitude" refers to the delimitation of a space which admits of no exception, as would be the case in a doctrine of virtue. In fact, Kant's reference to Hobbes is a constant throughout his work; it culminates in the 1793 text on the problematic relations at once conjoining and separating theory and practice, a text in which Kant develops a systematic critique of Hobbes.

As early as the *Critique of Pure Reason*, Kant agrees with Hobbes that there is injustice and violence in the state of nature; the coercion of law is necessary so that the coexistence of my freedom with that of others may be a lasting one (A752/B780). Which is the determining factor which pushes human beings to accept the passage into civil society? For Hobbes, it is violence itself as an empirical fact. Of course, Hobbes does wish to rise above the most ordinary experience: as he puts it in the *De Cive*, this would only show that what everyone wants from society is only either utility or the strictly egotistical pleasure of glory. The fact of society does not in any way follow from these facts; at most, they can account for spontaneous assemblies of people. An extended and lasting society can have no other origin than the mutual fear human beings develop with regard to one another. Mutual fear is a higher type of experience, for it is an "inference, made from the Passions."[22] The human capacity for reasoning is thus "deduced" from this passion, since one sees it confirmed by examples which show human beings *reflecting upon themselves*: " . . . when taking a journey, he armes himselfe . . . when going to sleep, he locks his dores; when even in his house, he locks his chests." Kant, however, does not allow himself to draw a concept from a passion or a feeling. The passage into civil society results instead from a conflict of feelings *within* the state of nature, a conflict which pushes man to move out of himself. In this regard, there is a remarkable congruence between the necessary condition for all knowledge and the properly human way of living in society; this congruence does not, however, indicate any historical evolution from one term to the other, but rather an inversion of their meaning. Recall that at the basis of all knowledge, according to critical philosophy, there lies reason's polemic with itself, in which hypotheses can be admitted for the momentary needs of a good cause. For the needs for defense, we can even completely invent our weapons, which are so many opinions we leave aside once defense is secured. The transmutation of dogmatic into critical philosophy must go through an antinomical conflict, which is a polemic that still takes itself seriously, but which then, because of the critique, turns into a war in which one fights only with leaden weapons. It is also the evolution from barbarism to civilization which is founded on an original conflict, but the order of the terms is reversed. The distinctive character of the human condition which leads to civil society is "a certain disingenuousness" which is none other than "a disposition to conceal our real sentiments, and to make show of certain assumed sentiments which are regarded as good and creditable." The search for an advantageous appearance (and thus for seriousness) has a civilizing effect because it keeps one from reflecting back upon oneself; no one, in fact, can "see through the outward show of respectability, honesty, and modesty" (A748/B776); I accept civil society because I take part in a game in which I seek to maintain an acceptable public appearance. Thus, as opposed to the case of knowledge, in public life falsehood wins out over truth.

Without the play of feelings that pushes private conscience outside of the subject by cloaking it in the external signs of morality, individual freedom could never be recognized as the basis for a perfect reciprocal coercion attuned to the freedom of all following universal external laws; from the perspective of right, my freedom would remain a freedom in name only, for it could never receive the sanction of a public law. In the *Metaphysics of Morals*, Kant goes one step further. The emergence from the state of nature is a commandment of reason, but one that would appear to have no basis other than the absolutely provisional acquisition of something external through occupation or contract. The civic state is comparable to the state of reflection accompanying action because it indicates "the conditions under which these laws are put into effect."[23] Though it is true that one cannot speak of a reciprocal and necessary relation between individuals before the establishment of a state of public legality, it is also true that one cannot speak of violence outside of civil society. It is as if it would be meaningless, from a transcendental perspective, to speak of violence reigning *blindly* in the state of nature. There, in accordance with an imperative of reason, all action always already moves in the *expectation* of the other's opinion. From the perspective of right, the fact of violence cannot be ascertained through immediate experience, because it only exists once when civil society has managed to reconstruct a completely ideal meaning for it, outside the sensible mediation of the other's opinion.[24] Thus, the way is closed to a necessary deduction of right in its universal principle, because it cannot do otherwise than to transform its requisite condition (the fact of violence in the state of nature) into an undesirable consequence of the exercise of its power in civil law. The concept here conforms to the fact, which it thereby no longer understands; there is no "Copernican" perspective which could make sense of the fact in conformity to a concept.

These considerations concerning the limits of a deduction of right converge with the fundamental themes of the *Critique of Judgment*. On the level of aesthetic experience, too, it is impossible to derive a concept from a solely private feeling. It is, of course, an imperative of reason that we leave the state of nature, but despite the absolute coercion this imperative places us under, its provisional foundation may yet waver. Such a precariousness of the absolute recalls the fundamental fragility of the aesthetic feeling, a fragility that is constantly covered over by the demand that the public necessarily assent to my private feeling. It is precisely only in the *aesthetic* feeling that one does not seek to exhibit feelings one does not have. It is perhaps the only kind of experience in which a priori conditions force us to live according to ideas. A reflection which aims at recovering the immediacy of contact with the singular is the inverse of the determining, temporalizing judgment because the latter already possesses the universal (concept or law). In the aesthetic feeling, we encounter an expectation that is nonetheless no longer dependent on time. Someone who says "this thing is beautiful" does not simply "count on other people to agree with his judgment on the ground that he has repeatedly found them agreeing with him; rather, he *demands* that they agree. He reproaches them if they judge differently, and denies that they have taste, which he nevertheless demands of them, as something they ought to have" (CJ 212–13). In the aesthetic judgment, ignorant of the challenge that my pretension so often meets with, I attribute to all of humanity beforehand a feeling that is identical to my own; the walls separating the public from the private are razed in the moment of feeling. The work of reflection consists in transforming a feeling which at first seems private and incommunicable into something communicable, through which I take others into consideration by putting myself in their place. Even though I take the others' points of view into consideration, it is only first to attribute to them my own feeling. I feel permitted to *do violence* to the other by encroaching on his judgment because through him I only ever address myself to all others, to the humanity which resides in him, also exceeds his person as well as mine. The other's judgment, on which I base my own feeling, is not a real judgment, but a merely possible judgment, without which the experienced conflict between judgments would never yield to any communication (CJ 293–94).

Doing violence to the other in this way (in the idea only, in the name of a humanity which is our common horizon) invariably results in a *feeling of violence* which must accompany every aesthetic feeling. There is in me a relationship to all of humanity, but it brings me no peace since it always exceeds me. Humanity serves as a universal (without which I could not judge at all), but one which is never initially guaranteed correspondence with the universal aimed at in reflection. This is because my aesthetic relationship with others is always only a possible relationship, not necessarily actual as in civil society. If it happens that there is even one person in the public who refutes my feeling, I undergo a shock that is all the more violent for the fact that I invest myself in my feeling, that I strive to live thereby in accordance with all of humanity. I thus find myself in direct conflict with morality, according to which I must seek to universalize the maxim of my action, in order to submit to the causality of beings in the intelligible world of freedom. Do not the initial demand and the subsequent reproach proceed from the same feeling of repulsion (which is the starting point for blame), namely, the repulsion before the always only possible character of my relation to others in terms of feelings, that is, the radical incompleteness of aesthetic concern? This "always only possible" is experienced as such even in its actualization, since nothing in the challenge put to my feeling stops me from trying again. This is the strange, provisional basis (human nature) for the absolute, for a legally organized human society. Are not blame and denial, then, as the vectors of aesthetic concern, the transcendental condition for violence? If this is correct, we can explain how it is that a certain violence persists in us (whether latent or expressed) even after the laws of freedom governing the external relations of people among themselves have been promulgated.

Let us now turn to the section from the 1793 text concerning theory and practice, which is presented as a systematic critique of Hobbes. Kant begins by rejecting the notion of the desire for happiness as the determinative principle of external right.[25] There is no determinative end, in fact, no determination at all; we are indeed within a reflective judgment. No determination is prior to man's capacity for reflection. This means that man is born adult; the time between the moment of his birth and the moment he reaches legally recognized maturity is of no consequence for right. Nothing about his birth makes him qualified to occupy a position of domination or to accept a servile condition.[26] Kant rejects any identifiable moment of origination for the social contract, and makes the case for an "*original contract*, among men on which alone a civil constitution can be based, one that is thus completely compatible with right and by means of which a commonwealth can be established."[27] Such a contract is a product of the general, unified will of the people. But once again it is meaningful only on the level of mere possibility. Indifferent to the actualization of the possibility it contains, the original contract is presented as a sort of aesthetic judgment in which my feeling seems to have managed to protect itself a priori, as it were, against any challenge from others. In demanding the consent of others, I no longer indulge in the fantasy of "someone abandoned on some desolate island" who is not content "with an object unless he can feel his liking for it in a community with others" (CJ 297). Not that this illusion is eradicated. Given the overwhelming task of living according to an idea of humanity which is never assured, the original contract allows one to experience the frustration that follows the repressed need for communication, that is to say, to suffer the paradox of solitude in the midst of society. Indeed, "if a public law is so formulated that an entire people *could not possibly* agree to it . . . it is not just; however, if *only* it is *possible* that a people could agree to it, it is a duty to regard that law as just."[28]

Since it remains within the domain of the possible, the original contract unconditionally prohibits any physically violent rebellion against the supreme legislative power. Violent opposition to the ruler presupposes a deceptive confusion between foundation and result, condition and conditioned. While the foundation may be certain, the result of such an action is always uncertain. Violence consists in a sort of contamination of the certain by the uncertain, or of the condition by the conditioned, such that the foundations become suspended, as it were, and the original contract gets abrogated: "might [instead of right] . . . renders all lawful constitutions insecure."[29] The cooperative subject, Kant writes, "must be able to assume that his ruler does not want to harm him."[30] One who legitimately considers himself a victim of power should understand the ruler's right to error and to ignorance of certain consequences of the laws. It is remarkable that Kant in no way considers the possibility of uncertainty concerning the intelligibility of the law itself. Why couldn't a radical uncertainty be the basis of both action and judgment? After all, Kant himself admits that it is "somewhat difficult to determine what requirements a person must meet so that he can be his own master."[31] If the amount of property necessary for citizenship is left indefinite, then it seems reasonable to claim that, despite himself, Kant invests the grounds of public right with a radical uncertainty. Only a naïve belief in the ruler's benevolence (which belief is not reflective in the least and accepts the play of determination) is capable of conferring an image of stability, one that is too easily taken for a reflection of the supposed certainty of the grounds of law.

As we might expect, what is at stake in the rejection of uncertainty with regard to the intelligibility of the laws is the overall consistency of the Kantian project. If we pursue the question of uncertainty, however, we push the concept of right into another domain. Let us return for a moment to Hobbes' conception of violence in the civil state. The mechanism of socialization is deduced from mutual fear, rather than from utility or egotism. This is essentially because fear is commutative and leads to the reciprocity of perspectives. This strict reciprocity, in its turn, leads "mechanically" to the explicit recognition of the general interest. Interest is distributed perfectly equally. But is this sufficient to eliminate all traces of uncertainty? Men's equality was, in practice, *already* the essential characteristic of the state of nature. Hobbes deduces war, as a phenomenon proper to the condition in the state of nature, from the equality of abilities and our equal hope of attaining our ends.[32] Thus, the passage into the civil state occurs without ever disturbing the principle of the equality of men. The transfer of power to the Leviathan can therefore have no other purpose than to make this passage visible. Since an individual is attributed rights and powers that place him infinitely above those who were his equals before the contract, there is no longer any cause for hostility between them and him, at least not for the hostility arising from the reciprocal presence of equal individuals. There can no longer be any question of natural violence between the sovereign and his subjects. But this system remains exposed to an entirely different kind of uncertainty, in connection with the function of visibility the sovereign fulfills. There is nothing to ensure the absence of another kind of violence: a political violence in the cases of abuse on the part of the head of state. One might then imagine another inference to be drawn from the passion for political violence; this inference would provide the basis for yet another contract designed to abrogate it, and so on.

This infinite regress is based on a radical uncertainty, one which is never really affected by the social contract, but merely displaced. One might wonder why Hobbes insists that the fundamental passion which makes it possible to overcome the divergence and conflict of opinions is the fear of dying *at the hands of others*. Hobbes considers the object of deliberation in the contract is to remove any just reason for fear. But isn't my life already vulnerable to a violent end in the state of nature? It seems that at the most basic level, the fear is a fear that man will stop nature from doing its work. A paradox ensues: in order to face up to the expectation of death and to preserve his natural end, man gives himself over to civil society, in which an infinitely powerful Leviathan utterly supplants nature. The fear is not lessened; it is, so to speak, buried alive in homage to inalterably all-powerful nature. Mathematical rigor in the metaphysics of the passions has not really rid this power of its sacred aura. Made visible and ruled by an infinite power, civil society escapes the control of its individual subjects and takes the place of nature. This substitution is successful and even desirable, because my own death escapes me in the same way that an infinite power never really exists. In addition to allowing peace between men, the infinite power they invent is the condition of possibility for nonhuman nature to carry out its work. Of course, we might be reminded, with Pascal, that man authentically measures up to himself when his fear of death is fear out of danger, not in danger. But isn't it also true, as Pascal goes on to argue, that nature makes do with the meagerest means to destroy him, for a vapor or a drop of water is enough to kill him?

Kant seeks precisely to cut short any series of subordinations that stretches to infinity.[33] The original contract, he writes, "[is] an infallible standard"[34] for determining whether the legislator is consistent with the principle of right. Kant states that the principle of infinite regress invalidates itself: "It would be an obvious contradiction for the constitution to include a law . . . giving the people the right to overthrow the existing constitution, from which all specific laws derive (even laws covering breach of contract), for it would then necessarily contain a *publicly constituted* opposing power."[35] Such a power would have its own leader, which would presuppose that at the origin of right one could postulate the existence of a lived contradiction between antagonistic parties who were always already there. If this were the case, then mutual coercion would have to exist *before* the recognition of finite domains for the exercise of freedom. The analytic principle of right, according to which coercion and finitude are simultaneous, would be ruined by this. To *live* within a contradiction is thus not at all the same thing as to *judge* according to its principle.

Now, the notion of humanity in the state of nature has no transcendental significance, since a minimum of culture must be posited in advance (CJ 355–56). For the peoples who experience a tendency toward legal society, there is always already an original difference between antagonistic parties, namely, a *social* difference between cultured and uncultured classes. Hence Kant's definition of humanity as a "universal feeling of sympathy" between the classes, which shows that the difference in question is prejudicial to the possibility of conceiving a reciprocal communication between what is innermost (the private judgment) and what is most universal (the idea of humanity which permeates every judgment of this kind). This communication would thus occur in the mode of taste. This is not to say that through taste a dissociation between the orders of nature and of culture arises at a certain moment in the history of a people, as if they became aware one day of the illusion they were living under while attached to the quasi-immutable determinations of nature. Not only does the illusion remain inasmuch as it is not detected at all (for we constantly distance ourselves from the state of nature, precisely because this state is unthinkable, resistant even to a retrospective determination), but it is also what allows us to open ourselves to the universal idea of humanity. Indeed, even when we become *completely* civilized, the work of reflection required to approach the idea will not be eliminated; as Kant points out, that pleasure which taste declares valid for mankind in general will remain. The indeterminable, original contradiction will always be with us, indefinitely repeated, to the same degree that taste cannot be enclosed in a concept and that its illusion cannot be detected.[36]

The bringing together of aesthetics and history is unique to Kant. Very little, if anything, remains of this fragile edifice in what Schelling did with it. Schelling passes from the spectator to the actor, whether this is the creative genius or the free agent of history. Interestingly enough, just as Heidegger interprets science as though it has succeeded, Schelling interprets the work of art as if it had succeeded. Art provides an *infinite* satisfaction, because in it all the opposites, whether conscious or unconscious, are eliminated, all contradiction is eliminated.[37] Once the product of art is completed, the artist feels an infinite satisfaction, as though the authentically successful work of art gave us, in anticipation, a foretaste of the end of history. What the product of art presents "is an infinite finitely displayed."[38] Faced with an authentically successful art work, no one would imagine anything better or different; thus it offers to intuition, as it were, a condensation of all history to come. Contrary to Kant's account, art is a power that separates a person from all others.[39] I do not, through art, desperately seek to link myself to all, because my freedom does not encounter the abyss of its finitude. To the contrary, the authentic artistic act begins with freedom; but when the infinity of the task threatens the work with unaccomplishment, a superior necessity takes over and allows for accomplishment. In that way, art appropriates for itself the sole future common to all of humanity. Is the artistic act, then, comparable to a moral act that contributes to the progress of humanity? In Kantian morality, a virtuous action manifests itself in the mode of the "ought to be," as a necessity that fills a blank in the actually sensible world of action. One can always hope for such progress, even if the history of humanity in its entirety had never displayed a single virtuous action. In accordance with Schelling, however, the marvel of art is that, "had it existed but once only, [it] would have convinced us of the absolute reality of that supreme event." An event, an example in the past, is necessary to enable art to fulfill its mission of the future. This absolute contingency is the price to be paid for "forcing" critical philosophy to speak the language of the transcendental future.

The culmination of this is that the singular event has so much weight that the entire possibility of intellectual intuition depends on it. Instead of referring to a matchless model to be imitated in some way, Schelling characterizes intellectual intuition in terms of the mind that places itself at the end of history—a point of indifference of thinking and intuiting. Schelling opposes himself virulently to the regressive synthesis that takes time, that is, the synthesis that takes the totality apart into cause and effect, because it stems from "the rage to explain everything, to accept nothing the way it is in its totality."[40]

In Kant's doctrine, right reduces an original difference within humanity as much as possible. This is a difference that the movement of history, and not simply (as for Hobbes) the movement of atomic individuals in brutish nature, designates as increasingly intolerable. In an ideally civilized society, the aesthetic feeling will always provide us with a sort of supplement of humanity, a difference between humanity and itself, because this feeling directs us back to the real purpose of the violence animating all conflicts, all differences—the origin of these differences is no longer even identifiable, for it is shrouded in the depths of the past. We find a fine example in the difference between the warrior and the statesman, a difference which Kant considers universally valid, as much for the primitive as for the most civilized of men. It is not the statesman but rather the general who inspires the greater admiration in the aesthetic sense. He is a man who is undaunted, who does not yield to danger, and whom one expects to "demonstrate all the virtues of peace," because he knows no fear and does not give way to danger, but sets manfully to work with full deliberation (CJ 262). In other words, since for him living and judging are one, he is alone in being able to claim that his reflection could pass for a determination. Such is not the case for the sovereign, for whom the right to error or ignorance removes any reason for being from the prestige of visibility. Therefore, the sovereign's function is rather to ensure the visibility of his subjects through the freedom of the public expression of opinions. This is why, in the final analysis, the only case in which rebellion is legitimate is one in which the freedom of expression has been abrogated.[41]

Does this imply that a knowledge (even if only ideal) of the universal bases of right allows any given public legality—provided it absolutely prohibits censorship—to avoid what Kant calls "a crime against human nature," namely, the implementation of an immutable contract requiring mutual obligation among its members? It lies in human nature's "proper destiny" to press the boundaries of knowledge ever further. Every age only ever lives in "expectation of a better [law]."[42] Now, the condition of possibility for such progress in knowledge is nothing other than the polemical use of reason, which consists in recognizing and utilizing the power contained in inevitable illusions. The critique of pure reason begins by neutralizing the dogmatic use of hypotheses, such as one finds in the antinomical conflict of pure reason. As for the polemical use of these hypotheses, which follows the critical solution of the antinomies, it gives us to understand that in a battle of this kind, both parties beat the air, and wrestle with their own shadows (A756/B784). Nonetheless, these bloodless conquests afford us with an indefinitely renewable pleasure because, even if the polemical use of hypotheses is limited to a game, we recall that, at any place or time, showing oneself to be better than one is is an essential component of civilization.

But doesn't the modern era present us with the image of polemics that become deadly games, like the French Revolution and its struggle to provide itself with a new beginning to history, one which would now be conscious of itself? Kant's ambiguity concerning this event is well-known. While he does consider violent revolutions to be wrong in every age, he cannot refrain from adopting a position which, as he puts it, "verges on enthusiasm."[43] This taking sides does not seem at all solid and secure; it gives one the impression of an aesthetic judgment unwilling to recognize itself as such. There is indeed a distance which keeps this position from taking the final step and admitting that it is simply aesthetic. Indeed, what gives the French Revolution its public significance is less the participants than the spectators who hail it without implicating themselves in it. In the case of this particular event, it is as if the judgment of the spectators was an aesthetic judgment which succeeded, as it were, precisely because it kept its distance from the place where history is made and unmade, that is, a place where I still have to fear being judged by others.

Two senses of the transcendental condition of possibility follow from this connection between knowledge and right. In the critique of pure theoretical reason, we move from facts to a legitimating concept, from the original finitude of the given to the aim of dialectical reason: the infinite. In the metaphysical doctrine of right, we begin with freedom as a given infinity in order to deduce finitude: the coexistence of finite domains of freedom embodied in the sensible world. The aesthetic judgment is the underlying pivot of these two movements. As we have seen in the case of revolution, where coexistence is put into question once more, the transcendental condition of possibility of *any* rebellion, legitimate or not, is the aesthetic judgment, which is always at least potentially open to being abrogated (through the other's challenge), and which is thus followed (at least ostensibly) by a rebellion against human nature. Regarding the aesthetic judgment, we may say that it always verges on a "crime against human nature" as this nature is given to us; this judgment always moves in the direction of a supposedly definitive modification of this nature according to new conventions. Only the general, by reminding us of our original condition of unresolved contradiction, could succeed in converting the frustration of a failed rebellion into pure satisfaction. This exception aside (and independently of our own doubts about its accuracy), life in civil society, just as the object of an aesthetic judgment, is the outcome of a disappointed expectation. After the logic of illusion, which could be overcome only by sinking to the bottom of it (the redoubled illusion), there comes the only possible reflection: that of delusion. And as far as moral philosophy is concerned, its chief problem is the following: we know the sense of the categorical imperative, within our innermost selves, but we have yet to know how such an imperative is possible. No example will help us, since the possibility in question would involve the transition to a viewpoint other than efficient causes; such a viewpoint would itself drag us into a wholly other realm of being, that is, the authentic self (*das eigentliche Selbst*) that thinks itself as practical. Practical reason would transcend its limits if it also willed to feel itself in the intelligible world. In the end, the critical philosopher, immersed in the stream of history and enamored of peace, must endeavor to think as if he could sense what cannot be sensed, on the border of the nonsensical.

We have seen that, according to Fichte, the present, in Kant's antinomies, is just as unconditioned as the most remote past. In what sense? Shall we speak of present pregnant with the whole future? No: for Kant, there is indeed a sense of the unconditioned for the present, but as the perilous self-maintenance of life. Nietzsche certainly did not think that he epitomized so well the whole of critical philosophy, when he exclaimed "Will to truth—that might be a concealed will to death."[44] My taste, which is also my feeling of life, according to the first paragraph of the *Critique of Judgment*, is constantly revoked; and yet it is necessary to go on living. There remains for us only finitude. Of the two Heideggerian senses of finitude (project for being and dependence on beings as pregiven), let us focus on the second. This dependency ends in outright submission: "In order to allow the being to be what and as it is, . . . the existing being [Dasein] must already have projected that it is a being on the strength of what has been encountered. Existence means dependency upon the being as such in the submittance to the being as such which is dependent in this way."[45] In working out a cosmological principle in Kant's Dialectic, we have seen that the strength of what has been encountered comes up against a layer of being in which Dasein has not always already projected itself. This layer of being is the illusion that comes to be for the human thinking creature—what Kant calls the unobserved influence of sensibility upon the understanding.

Heidegger asks how the question of Being will ever be meaningful, if it remains obscure as to from whence we can expect an answer. The direction for an answer will be provided by what Heidegger calls the preconceptual understanding of Being—that which we constantly understand, always and already, must be brought to the surface. The primacy of the future in the existential analytic is then easily accounted for, since only the anticipation of death (the exhaustion of all that we have already understood) can unfold the proper sense of temporality for a finite being. Any ontological comprehension, Heidegger says, is inevitably construction. But construction does not mean free invention. Rather, it means "a projecting in which the preliminary guidance as well as the taking-off of the projection must be predetermined and protected."[46] By completing Kant's project, by taking the starting point of this project in the arbitrariness of the absolute which is not yet human (even though it resides in the innermost depths of our lives), Fichte already reminds us that there is no predetermination, at any moment of the existence of a being, that does not begin to waver once it attempts to relate to itself. Kant's own radicality goes perhaps even further. Reflective judgment allows us to oppose a sense of guidance, different from Heidegger's, for the predetermination of Being: the preconceptual is simply not a deeper layer of the concept, but rather its back. The preconceptual understanding which guides us in the labyrinth of rules and pseudo-rules, true and false appearances, could well be entirely outside the concept. In the final analysis, we communicate with the layer of being in which human being has not always already projected itself only through taste, and still in the mode of crushing illusion.

Thus, taste gives us death by anticipation. This explains why the future is not and cannot be a transcendental problem. Death is not anticipated by Dasein's authentic relation to it; rather, death is given to it at every moment of aesthetic experience—the absolute of cognition. At such a moment, death has to be lived—I shall never know how, I shall never be able to determine whether this experience is authentic or not.

As far as theory is concerned, the pillar of Kantian reason is the postponing of reason's fulfillment in being. The suspension never really terminates, and in this sense knowledge retains the classical movement of erotic ascent that Kant has inherited from the Platonic tradition. Postponing requires the impossible totality of a temporal synthesis. But in teleology, Kant compels us to think in terms of substitution, because judgment displaces or dislocates the fulfillment of being. Space, as the total form of experience, now remains the only mode in which our discursive understanding can think in the manner of a superior intellect. As Kant has shown in his example of the dark, yet familiar, room, one of our innermost feelings (the subjective sense of left and right) can slip, through the mediation of everyday experience, into the pure intuition of space. It is the violence of the everyday world in its alternation between day and night that ultimately forces our outer sense into another world. Likewise, and even more profoundly, the experience of aesthetic estimation is a sublime and painful one; in order to comprehend a multiplicity in one instant, rather than in accordance with the temporal succession in objective measuring, the imagination must do violence to inner sense; the violence and pain increase in proportion to the greatness of the quantum that has to be comprehended in one intuition. The violence of the aesthetic absolute reduces inner sense to silence, so that finally, through this self-inflicted pain, our understanding is reminded of its supersensible vocation. We never begin in comforting praxis and then tranquilly raise ourselves up into theory or ontology. Whether it is a totality of space or whether it is a totality of time, from the start something that occurs outside me always assails me within myself.

Notes

1. Totality, Finitude, and Division

1. *Critique of Pure Reason*, trans. N. K. Smith (London: Macmillan, 1929), p. 657 (A838/B866). It will be quoted hereafter in the text in accordance with the standard page numbers of the standard German editions, A and B. All works by Kant other than the *Critique of Pure Reason* and the *Critique of Judgment* (see below, note 2) will be referred to in notes and will follow the standard "Akademie Ausgabe" (Ak). The roman numeral that follows "Ak" indicates the volume number of that standard edition.

2. *Logic*, Ak, pp. 23–25; *Critique of Judgment*, sec. 9 (Ak.V, p. 316) and sec. 83 (Ak.V, p. 432). This latter work will be quoted hereafter in the text as CJ, followed by the page numbers of the standard German edition. These numbers are indicated in the English translation that we will use by W. Pluhar (Indianapolis: Hackett, 1987).

3. M. Heidegger, *The Basic Problems of Phenomenology*, trans. A. Hofstadter (Bloomington: Indiana University Press, 1982), pp. 7–12 and 49ff.

4. M. Heidegger, *Kant and the Problem of Metaphysics*, trans. R. Taft (Bloomington: Indiana University Press, 4th ed. 1990), p. 8.

5. Ibid., p. 7.

6. The critical starting-point for any evaluation of the role of Kant's Dialectic in relation to Heidegger is provided by H. Declève, *Heidegger et Kant* (The Hague: M. Nijhoff, 1970), pp. 331–70.

7. G. W. F. Hegel, *Vorlesungen über die Geschichte der Philosophie*, ed. Glockner, vol. 17, p. 343.

8. F. Nietzsche, *Der Wille zur Macht*, sec. 556, p. 381.

9. Ibid., sec. 544, p. 370.

10. E. Cassirer, "Kant and the Problem of Metaphysics," trans. M. S. Gram in *Kant: Disputed Questions* (Chicago: Quadrangle Books, 1967), pp. 146–47.

11. *On the Dignified Tone Recently Adopted in Philosophy*, Ak.VIII, pp. 398–99.

12. *Critique of Practical Reason*, Ak.V, p. 161.

13. I. Kant, "What is Orientation in Thinking?", Ak. VIII, p. 135.

14. M. Heidegger, *Being and Time*, sec. 23, trans. J. Macquarrie and E. Robinson (New York: Harper & Row, 1962), p. 144. Already within the year that followed the publication of "What is Orientation in Thinking?" Wizenmann (in *Deutsches Museum*, February 1787, pp. 132–3) levelled a similar objection against Kant. He argued that, if indeed we are to make a correct judgment on the position of not yet visible objects, then the subjective principle of differentiation between right and left must be complemented by an objective principle of differentiation, namely, the very object that I seize.

15. "What is Orientation in Thinking?", p. 136.

16. *Being and Time*, p. 234.

17. F. W. J. Schelling, *System of Transcendental Idealism*, trans. P. Heath (Charlottesville: University Press of Virginia, 1978), p. 26.

18. M. Heidegger, *Schelling's Treatise on the Essence of Human Freedom*, trans. J. Stambaugh (Athens, OH: Ohio University Press, 1985), p. 38.

19. M.Heidegger, *Vom Wesen der menschlichen Freiheit* (1930), Gesamtausgabe vol. 31 (Frankfurt am Main: Klostermann, 1982), pp. 214–15.

20. Ibid., p. 226.

21. Ibid., p. 246.

22. Ibid., pp. 239–44.

23. M. Heidegger, *Hegel's Concept of Experience*, trans. K. R. Dove (New York: Harper & Row, 1970), p. 37.

24. E. Cassirer, "Kant and the Problem of Metaphysics," p. 142.

25. *Kant and the Problem of Metaphysics*, p. 147.

2. The Mathematical Dream of Philosophy

1. M. Heidegger, "Who is Nietzsche's Zarathustra?," orig. in *Vorträge und Aufsätze* [1954], trans. B. Magnus, *Review of Metaphysics*, 20 (1967): 411–31.

2. *On the Dignified Tone*, p. 398.

3. *Anthropology from a Pragmatic Viewpoint*, sec. 52 (Ak. VII, p. 215).

4. These examples are given in *Critique of Pure Reason*, p. 180 (A137/B176). There is no need to correct Kant's sentence, as suggested by Vaihinger and accepted by Kemp-Smith, which would then read: what is necessary to think the roundness in a circle, or to intuit this roundness in a plate. Indeed, Kant's point is precisely to *oppose* intuition and thought in order to show what the capacities of schematism are.

5. *Anthropology*, sec. 53, Ak. VII, p. 219.

6. Ibid., sec. 2, Ak. VII, p. 129.

7. This point is emphasized in *Idea for a Universal History from a Cosmopolitan Point of View*, trans. L. W. Beck in *Kant on History* (Indianapolis: Bobbs-Merrill, 1963), pp. 11–12 (Ak. VIII, p. 17).

8. *Dreams of a Spirit-Seer*, Ak. II, p. 348.

9. Galileo, "The Assayer," in *Discoveries and Opinions of Galileo*, trans. S. Drake (New York: Doubleday, 1957), p. 337.

10. Ibid., pp. 237–38.

11. See J. Heilbron, *Elements of Early Modern Physics* (Berkeley: University of California Press, 1982), pp. 2–3.

12. "The Assayer," p. 238.

13. *On the Dignified Tone*, p. 405.

14. Uniqueness and infinity are the third and the fourth elements of the metaphysical exposition of space (see A25/B39–40); they are the fourth and fifth elements of the metaphysical exposition of time (see A31–32/B47).

15. M. Heidegger, *Kant and the Problem of Metaphysics*, p. 148.

16. Ibid., p. 31.

17. Ibid., p. 16.

18. Ibid., p. 20.

19. See Kant's remarkable essay of 1794, "The End of all Things", Ak. VIII, p. 334.

20. Kant writes that "the ideal is . . . the archetype (*prototypon*) of all things, which one and all, as imperfect copies (*ectypa*), derive from it the material of their possibility, and while approximating to it in varying degrees, yet always fall very far short of actually attaining it."

21. Heidegger disregards space as pure image and focuses essentially on time. Consequently, he argues that "time is not only the necessary pure image of the schemata of the pure concepts of the understanding, but also their sole, pure possibility of having a certain look" (*Kant and the Problem of Metaphysics*, p. 71). I contend that there can be no pure image of a schema. Moreover, that time is the pure image of *all* schemata is certainly not supported by Kant's text, since Kant refrains from ascribing any image whatsoever to the categories other than quantity.

22. See H. Jonas, "Is God a Mathematician?," in *The Phenomenon of Life* (New York: Harper& Row, 1966), pp. 64–98, and J. Klein, "The World of Physics and the 'Natural' World," trans. D. Lachterman, in W. Williamson and E. Zuckerman (eds.), *The Lectures and Essays of Jacob Klein* (Annapolis: St John's College Press, 1985), pp.1–34.

23. *Religion within the Limits of Reason Alone*, Ak. VI, p. 25. "Innate constitution" translates "angeboren Beschaffenheit." Kemp-Smith's rendering of *Gesinnung* by "sentiment" is particularly weak.

24. *Metaphysical Foundations of Natural Science*, trans. J. Ellington (Indianapolis: Hackett Publishing Company, 1985), p. 27 (Ak. IV, pp. 486–87). Quoted hereafter as MFNS.

25. MFNS, pp. 40–41 (Ak. IV, pp. 496–97).

26. Cf. J. Vuillemin, *Physique et Métaphysique Kantiennes* (Paris: Presses Universitaires de France, 1955), pp. 92–93.

27. MFNS, pp. 65–66 (Ak. VI, p. 515).

28. Ibid., pp. 92–93 (Ak. VI, p. 534).

29. Ibid., pp. 58–59 (Ak. VI, p. 510).

30. Ibid., pp. 42–43 (Ak. VI, p. 498).

31. B. Pascal, "Disproportion of Man," *Pensées*.

32. E. Husserl, *The Crisis of European Sciences and Transcendental Phenomenology*, trans. D. Carr (Evanston: Northwestern University Press, 1970), pp. 112, 143.

33. E. Husserl, *Erste Philosophie*, Zweiter Teil: "Theorie der phänomenologischen Reduktion," in *Husserliana*, vol. 8 (The Hague: M. Nijhoff, 1959), pp. 41–42.

34. *Prolegomena to Any Future Metaphysics*, trans. J. Ellington (Indianapolis: Hackett Publishing Company, 1985), p. 58 (Ak. IV, p. 315); translation slightly modified.

3. An Experiment with Concepts

1. *Prolegomena*, p. 93 (Ak. IV, p. 352).

2. Ibid., pp. 97–98 (Ak. IV, p. 357).

3. B. Russell, *Human Knowledge* (New York: Simon and Schuster, 1948), p. 9.

4. N. R. Hanson, "Copernicus' Role in Kant's Revolution," *Journal of the History of Ideas* 20 (1959): 274–81.

5. N. Capaldi, "The Copernican Revolution in Hume and Kant," in L. W. Beck (ed.), *Proceedings of the Third International Kant Congress* (Dordrecht: Reidel, 1972), p. 239.

6. On the significance of history for pure reason, see J. Sallis, *Spacings of Reason and Imagination* (Chicago: University of Chicago Press, 1987), pp.1–8.

7. *Prolegomena*, p. 19 (Ak. IV, p. 274).

8. H. Blumenberg, *The Genesis of the Copernican World*, trans. R.Wallace (Cambridge, MA: MIT Press, 1987), p. 609.

9. MFNS, pp. 67–68 (Ak. IV, pp. 516–17).

10. Ibid., p. 62 (Ak. IV, p. 513).

11. I. Newton, *Mathematical Principles of Natural Philosophy* (1726), trans. A. Motte, rev. F. Cajori (Berkeley: University of California Press, 1934), p. 419. Quoted hereafter as *Principia*.

12. Ibid., p. 544.

13. Voltaire, *The Elements of Sir Isaac Newton's Philosophy*, trans. J. Hanna (London, 1738), p. 348.

14. Ibid., pp. 345–46.

15. "Four letters of Sir Isaac Newton to Doctor Bentley," in *Isaac Newton's Papers and Letters on Natural Philosophy*, eds. I. B. Cohen and R. E. Schofield (Cambridge, MA: Harvard University Press, 1958), pp. 305–6.

16. *Prolegomena*, p. 93 (Ak. IV, p. 352). Further emphasis on the infinite progress of these sciences is found in MFNS, p. 11 (Ak. IV, p. 473).

17. *Prolegomena*, p. 94 (Ak. IV, p. 354).

18. Ibid., p. 96 (Ak. IV, p. 355).

19. "MFNS, p. 120 (Ak. IV, p. 555). Very early in his career (1747), Kant had noted that "motion is the external phenomenon of force, but the tendency to conserve motion is the basis of all activity" (*Thoughts on the True Evaluation of Living Forces*, Ak. I, p. 141). As it turns out, the combination of reason and understanding is not only running up against sensibility as a cognitive faculty, but also sensibility as the seat of aesthetic judgments (CJ 222). The consciousness that accompanies a purely contemplative pleasure contains a basis for the "quickening" of a subject's cognitive powers. This quickening is an inner causality consisting in "*keeping* [us in] the state of [having] the presentation itself. . . . We *linger* in our contemplation of the beautiful, because this contemplation reinforces and reproduces itself.

20. See M. Prieur, "En quoi, comment et pourquoi la dialectique est, pour Kant, une logique de l'apparence?," *Actes du XIVème Congrès des Sociétés de Philosophie de Langue Française* (Nice, 1969), pp. 92–95.

21. *Universal History of Nature and Theory of the Heavens*, trans. S. Jaki (Edinburgh: Scottish Academic Press, 1981), p. 152. Without he could possibly know it, Kant revives here one of Newton's arguments in his letters to Bentley. See ch. 5 for how Newton discards this fear of slightest displacement.

22. Ibid., p.104.

23. Ibid., p.154.

24. *Critique of Practical Reason,* trans. L. W. Beck (Chicago: University of Chicago Press, 1949), p. 258 (Ak. V, p. 161).

25. "Uber die Vulcane im Monde," *Berlinische Monatschrift* 5 (1785): 199–213. English translation by O. Reinhardt and D. Oldroyd in "By analogy with the heavens: Kant's theory of the earth," *Annals of Science* 41 (1984): 221.

26. *Opus Postumum,* Ak. XXI, p. 97.

27. See the commentary by M. Prieur, "La Référence Optique dans la Topique chez Kant," *Archives de Philosophie* 35 (1972): 372.

28. *Anthropology,* Ak. VII, p. 137 note).

29. *Prolegomena,* p. 100 (Ak. IV, p. 360).

30. I. Newton, *Principia,* pp. 614–15.

31. F. Bacon, *Novum Organum,* Aphorisms 41 and 47. (See also aphorisms 69 and 115.)

32. G. Galileo, *Dialogue Concerning the Two Chief World Systems,* trans. S. Drake (Berkeley: University of California Press, 1967), p. 256.

33. See H. Arendt, *The Human Condition* (Chicago: University Press of Chicago, 1958), p. 63.

34. MFNS, p. 9 (Ak. IV, p. 472).

35. *Idea for a Universal History from a Cosmopolitan Point of View,* p. 22 (Ak. VIII, p. 27).

36. Ibid., p. 23 (Ak. VIII, p. 28).

4. Reversing the Order of Time

1. I. Newton, *Principia,* p. 13.

2. Ibid., p. 6.

3. The words "in general," a crucial qualification of "something else," have curiously been omitted in Kemp-Smith's translation.

4. *Correspondence,* Ak. X, p. 269.

5. *Prolegomena,* p. 79 (Ak. IV, p. 338).

6. *Correspondence,* Ak. X, p. 252.

7. Quoted in L. W. Beck, *Early German Philosophy* (Cambridge, MA: Belknap Press, 1969), p. 462.

8. *Prolegomena,* p. 81 (Ak. IV, p. 341).

9. Heidegger seems to have missed the distinction between the natural and the lawful. In his interpretation, the two are conflated because he speaks of "a natural conflict of the laws of pure reason" *(Vom Wesen der menschlichen Freiheit*, p. 210).

10. Galileo, "Letter to the Grand Duchess Christina," in *Discoveries and Opinions of Galileo*, p. 202.

11. See in particular K. Hübner, *Critique of Scientific Reason*, trans. P. Dixon and H. Dixon (Chicago: Chicago University Press, 1983), pp. 143–44.

12. The point is made by G. Whitrow, "On the Impossibility of an Infinite Past," *British Journal for the Philosophy of Science*, 29 (1978): 41.

13. See for instance J. Bennett, "The Age and Size of the World," *Synthese* 23 (1971): 127–46.

14. See for instance S. Al-Azm, *The Origins of Kant's Arguments in the Antinomies* (Oxford: Clarendon Press, 1972); H. Allison, *Kant's Transcendental Idealism* (New Haven: Yale University Press, 1983).

15. H. G. Alexander (ed.), *The Leibniz-Clarke Correspondence* (Manchester: Manchester University Press, 1956), p. 34.

16. J. Merleau-Ponty and B. Morando speak of classical cosmology as a "pseudo-cosmology" in *The Rebirth of Cosmology*, trans. H. Weaver (New York: A. Knopf, 1976), p. 67.

17. The charge of insidiousness on the part of Kant is now a commonplace: thus Deleuze writes that "if there were only the speculative interest, it would be very doubtful whether reason would ever consider things in themselves," in *Kant's Critical Philosophy*, trans. H. Tomlinson and B. Habberjam (Minneapolis: University of Minnesota Press, 1984), p. 6.

18. "What is Orientation in Thinking?", Ak. VIII, p. 142.

19. Ibid., p. 141.

20. Attempts of this sort are presented in L. Schäfer, "The Restrictive and Proliferative Function of Kant's Regulative Ideas," in L. Beck (ed.), *Proceedings of the Third International Kant Congress* (Dordrecht: Reidel, 1972), pp. 480–85; S. Toulmin, *The Return to Cosmology: Postmodern Science and the Theology of Nature (*Berkeley: University of California Press, 1982), pp. 3–8.

21. W. Werkmeister, "Kant's Philosophy and Modern Physics," in W.Werkmeister (ed.), *Reflections on Kant's Philosophy* (Gainesville: University Presses of Florida, 1975), pp. 109–33. See p. 118 in particular.

22. In connection with the third antinomy, F. Alquié writes that "Kant, tout en raisonnant encore dans la perspective des dogmatistes, prépare la voie de sa propre théorie," in *La Critique Kantienne de la Métaphysique* (Paris: Presses Universitaires de France, 1968), p. 81.

23. See for instance R. Swinburne, *Space and Time* (New York: St. Martin's Press, 1968), pp. 280ff.

24. C. D. Broad, "Kant's Mathematical Antinomies," *Proceedings of the Aristotelian Society* 55 (1954–55): 2–20.

5. A Logic of Illusion

1. M. Heidegger, *What is a Thing?*, trans. W. B. Barton and V. Deutsch (Chicago: Regnery/Gateway, 1967), p. 20.

2. Ibid., p. 14.

3. A. S. Eddington, *The Nature of the Physical World* (Cambridge: Cambridge University Press, 1928), pp. xi–xix.

4. *Isaac Newton's Papers and Letters on Natural Philosophy*, p. 281.

5. *The Leibniz-Clarke Correspondence*, pp. 31, 46.

6. *Isaac Newton's Papers and Letters on Natural Philosophy*, pp. 294–96.

7. *The Leibniz-Clarke Correspondence*, pp. 75–78.

8. "Letter to Bourget," 5 August 1715, in *Leibniz: Philosophical Papers and Letters*, trans. L. Loemker (Chicago: Chicago University Press, 1956), vol. 2, p. 1080.

9. See Simplicius, *De Caelo*, 303.33 sqq.

10. Aristotle, *On the Heavens*, 280a4sq.

11. Plotinus, *Enneads*, 3.5.24sq.

12. Plato, *Timaeus*, 48d.

13. *Le Monde*, in *Descartes: Oeuvres complètes*, ed. Ch. Adam and P. Tannery (Paris: Vrin, 1974), p. 31.

14. *Prolegomena*, p. 101 (Ak. IV, p. 361).

15. *Critique of Practical Reason*, Ak. V, pp. 31, 35.

16. Ibid., p. 69.

17. Galileo, "Letter to the Grand Duchess Christina", p. 182.

18. *Isaac Newton's Papers and Letters on Natural Philosophy*, p. 303.

19. M. Heidegger, *Vom Wesen der menschlichen Freiheit*, p. 213.

20. See, among many possible primary sources, E. Husserl, *Phenomenological Psychology*, trans. J. Scanlon (The Hague: M. Nijhoff, 1977), p. 55.

21. *Opus Postumum*, Ak. XX, p. 339.

22. MFNS, p. 20 (Ak. IV, p. 481).

23. Hegel, *Wissenschaft der Logik*, ed. G. Lasson (Hamburg: Meiner, 1932, vol.1), p. 183.

24. H. Graubner (*Form und Wesen*, Kant-Studien Ergänzungshefte 104, Bonn, 1972, pp.147–65) argues that the absence of concepts in the unity of formal sensibility proves the existence of a precategorial influence of the apperception on pure intuiting. It would seem that an even more radical approach is desirable: a formal intuition has a unity that is categorial in its own way, that is, sensibility does not even borrow its unifying function from the understanding. Cf. our remarks above concerning space as a thinking intuition. A complete clarification of these issues is provided by A. Ferrarin, "Construction and Mathematical Schematism. Kant on the Exhibition of a Concept in Intuition," *Kant-Studien* 86 (1995): 131–174.

25. C. D. Broad, "Kant's Mathematical Antinomies", pp. 5, 11–12.

26. Such a perceptive commentator as Schopenhauer could recognize only a "happy coincidence" in the fact that Kant could apply the category of quantity to the problem of the limits of the world. Our expression "absolute quantity" is itself suggested by a historical circumstance. With regard to the possible parallel between mathematical and cosmic genesis, Leibniz writes that "we can . . . understand in a wonderful way how a kind of divine mathematics or metaphysical mechanism is used in the origin of things and how the determination of the maximum [greatest quantity] takes place." ("On the Radical Origination of Things," in *Leibniz: Philosophical Papers and Letters*, p. 792.)

27. *An Attempt to Introduce in Philosophy the Concept of Negative Magnitudes*, Ak. II, p. 169.

28. I. Newton, *Opticks*, 4th ed. (New York: Dover), question 31, p. 395.

29. *An Attempt*, Ak. II, p. 197.

30. See another work of Kant's work of the same year 1763: *The One Possible Basis for a Demonstration of the Existence of God*, trans. G. Treash (New York, 1979), Ak. II, pp. 77–78.

31. A. Schopenhauer, *The World as Will and Representation*, trans. E. F. J. Payne (New York: Dover, 1969), vol. 1, p. 493.

32. *Prolegomena*, p. 94 (Ak. IV, p. 354; translation modified). N. Rescher draws the useful distinction between boundaries as "terminating limits" and limits as "excluding limits." ("Kant on the Epistemology of Scientific Questions," in *Kant's Theory of Knowledge and Reality* [Washington, DC: University Press of America, 1983], p. 68.)

33. I. Kant, *What Real Progress has Metaphysics Made in Germany Since the Time of Leibniz and Wolff?*, trans. T. Humphrey (New York: Abaris Books, Inc., 1983), p. 117 (Ak. XX, p. 291).

34. M. Heidegger, *Schelling's Treatise on the Essence of Human Freedom*, p. 41.

35. *Kant and the Problem of Metaphysics*, p. 146.

6. A Reversal of the Reversal

1. Hegel, *Wissenschaft der Logik*, p. 184. See also M. Guéroult's commentary, "Le Jugement de Hegel sur l'Antithétique de la Raison Pure," *Revue de Métaphysique et de Morale*, 1931, pp. 412–39.

2. M. Heidegger, *Schelling's Treatise on the Essence of Human Freedom*, p. 58.

3. F. W. J. Schelling, *Ideas on a Philosophy of Nature as an Introduction to the Study of this Science*, 2nd ed. 1803, trans. P. Hayden-Roy in *Philosophy of German Idealism*, ed. E. Behler (New York: Continuum, 1987), p. 169.

4. See Kant's letter to Marcus Herz, 26 May 1789, in *Kant: Philosophical Correspondence*, trans. A. Zweig (Chicago: University of Chicago Press, 1967), p. 151 (Ak. XI, 2, p. 49).

5. S. Maimon, *Über die Progressen der Philosophie* (Berlin: 1793), p. 20.

6. S. Maimon, *Versuch über die Transzendentalphilosophie* (Berlin: 1790), in *Gesammelte Werke*, vol. 2, ed. V. Verra (Hildesheim: G. Olms, 1965–76), p. 50.

7. Ibid., p. 350.

8. See J. Vuillemin, *Physique et Métaphysique Kantiennes*, p. 52.

9. J. G. Fichte, *Grundriß des Eigentümlichen der Wissenschaftslehre* (1795), Sämmtliche Werke (S.W.)1, p. 362.

10. J. G. Fichte, "First Introduction to the Science of Knowledge," in *The Science of Knowledge*, trans. P. Heath and J. Lachs (Cambridge: Cambridge University Press, 1984), p. 26.

11. *The Science of Knowledge*, p. 99.

12. "First Introduction", p. 8.

13. *The Science of Knowledge*, p. 104.

14. Ibid., p. 200 (transl. modified).

15. *Nachgelassene Schriften*, vol. 2, p. 299.

16. *The Science of Knowledge*, p. 217.

17. *Opus Postumum*, Ak. XXI, pp. 4, 6.

18. Platon, *Theaetetus*, 179.

19. M. Heidegger, *Vom Wesen der menschlichen Freiheit*, p. 213.

20. *The Conflict of the Faculties*, Ak. VII, pp. 83–84.

21. *The Science of Knowledge*, p. 112.

22. Ibid., p. 116.

23. I. Kant, *The End of All Things*, Ak. VIII, p. 334.

24. *Grundriß*, S.W.1, p. 410.

25. *The Science of Knowledge*, p. 171.

26. Emphasis added.

27. *Opus Postumum*, Ak. XXI, p. 211.

28. Ibid., Ak. XXI, p. 214.

29. Ibid., p. 210.

30. Ibid., Ak. XX, p. 215.

31. Ibid., Ak. XXI, p. 566.

32. Ibid., Ak. XXII, p. 548.

33. MFNS, p. 7 (Ak. IV, pp. 470–71).

34. A note of Kant's *Nachlass*, which according to Adickes dates from the same year 1789, speaks about the possibility of chemistry being an apodeictically grounded science (Ak. XIV, pp. 489, 492).

35. The latter view has been defended by M.Friedman, *Kant and the Exact Sciences* (Cambridge, MA: Harvard University Press, 1992).

36. Kant also deals with the idea of soul, which must be viewed as if it were a real being. If I enquire about the soul in itself, I "abstract . . . from all conditions requisite for thinking an object for such a concept; yet only as related to an object can the concept be said to have meaning" (A684/B712).

37. At the beginning of section 70, Kant states more simply that "reason deals with nature as the sum total of objects of outer senses" (CJ 386).

38. This is the basis of Hegel's argument in *Wissenschaft der Logik*, vol. 2, p. 387ff.

39. Emphasis added.

40. The suggestion is discussed by A. Philonenko, "L'antinomie du jugement téléologique chez Kant," *Revue de Métaphysique et de Morale*, 82 (1977): 13–37.

41. Two different interpretations of the solution should be critized. Butts (*Kant and the Double Government Methodology*, Dordrecht: Reidel, 1984, pp. 272–73), among others (e.g., Philonenko), argues that the antinomy is solved by transforming the constitutive back into the regulative. This interpretation overlooks the fact that

there is already a conflict between the two maxims, which is itself not solved by the "solution" of the antinomy. By contrast, Pluhar, in his introduction to the translation of the *Critique of Judgment*, tends to overestimate the conflict. He interprets Kant as saying that "we are contradicting ourselves unless we can reconcile the two principles" (p. xc). But there is no point in reconciling them, because only the two taken together constitute the unconditioned, the demand of which has already been satisfied. Reflection accommodates itself with a conflict which, after all, is no more than seeming. (In an antinomy of reason, the conflict is not merely seeming, in the sense that it remains, even after it has been detected by an appropriate critique.)

42. In his detailed commentary of the antinomy of teleological judgment, L. Ferry (*Philosophie Politique*, vol. 2, Paris: Presses Universitaires de France, 1984, pp. 210–42; see p. 221 in particular) maintains the traditional duality between dogmatism and critique, inasmuch as the former cannot ground the viewpoint from which it attempts to prove its assertions. But if the same process is now characterized in terms of the "blind spot" of the entire critical enterprise, then a very different interpretation of the antinomy becomes possible.

43. Kant writes that, if the principle is determinative, "it strays beyond the world of sense and into the transcendent, and perhaps will be misled" (CJ 389).

44. See E. Sober, *The Nature of Selection: Evolutionary Theory in Philosophical Focus* (Cambridge, MA: MIT Press, 1984), ch.2.

45. Second emphasis is added.

46. The latter alternative is the cosmogonical model of explanation used by Kant in his pre-critical *Theory of the Heavens*.

47. A. Einstein, *The Meaning of Relativity*, trans. E. P. Adams (London: Chapman and Hall, 1967), p. 2.

48. *Ueber den Begriff der Wissenschaftslehre*, Gesamtausgabe der Bayerischen Akademie den Wissenschaften (G.A.),1, p. 143.

49. Ibid., p. 146.

50. *The Science of Knowledge*, p. 189.

51. Ibid., p. 243.

52. Ibid., p. 230.

53. Ibid., p. 258.

54. Ibid., p. 259.

55. Ibid., p. 261.

56. Ibid., p. 262.

57. Ibid., p. 276.

58. Ibid., p. 277.

59. Ibid., p. 280.

60. Ibid., p. 283.

61. *Theaetetus*, 201d.

62. This analogy between Fichte's absolute self and one's own body is suggested by what Merleau-Ponty has to say about the "corps propre" in *La Structure du Comportement* (Paris: Presses Universitaires de France, 1942), pp. 230–31.

7. Lost Illusions

1. "First Introduction," in *The Science of Knowledge*, p. 12.

2. S.W. IV, p. 50.

3. *Idea for a Universal History from a Cosmopolitan Point of View*, p. 15 (Ak. VIII, p. 20).

4. Ibid., pp. 12–13 (Ak.VIII, pp. 18–19). The prototype of such an incomplete development in the case of cognition was transcendental schematism.

5. Ibid., p. 15 (Ak. VIII, p. 21), translation modified.

6. *Metaphysics of Morals*, "Doctrine of Right," Ak. VI, p. 350.

7. S. Maimon, *Versuch über die Transzendentalphilosophie*, p. 338 (emphasis added).

8. M. Heidegger, *Phänomenologische Interpretation von Kants Kritik der reinen Vernunft* (Frankfurt am Main: Klostermann, 1977), p. 410.

9. *Kant and the Problem of Metaphysics*, p. 161.

10. Ibid., p. 156.

11. *Being and Time*, pp. 436–37.

12. *The Metaphysics of Morals*, trans. Mary Gregor (New York: Cambridge University Press, 1991), p. 57 (Ak. VI, p. 231).

13. Ibid., p. 69 (Ak. VI, p. 247).

14. Ibid., pp. 56–57 (Ak. VI, p. 231).

15. Ibid., p. 76 (Ak.VI, p. 255).

16. *Critique of Practical Reason*, Ak. V, p. 46.

17. *Metaphysics of Morals*, p. 58 (Ak. VI, p. 232).

18. Ibid., p. 83 (Ak. VI, p. 262).

19. "On the Proverb: That May be True in Theory, But Is of No Practical Use" (hereafter "Theory and Practice"), in *Perpetual Peace and Other Essays*, trans. Ted Humphrey (Indianapolis: Hackett, 1983), p. 77 (Ak. VIII, p. 297).

20. See L. Strauss, *The Political Philosophy of Hobbes* (London: Oxford, 1936).

21. *Metaphysics of Morals*, p. 58 (Ak. VI, p. 233).

22. Th. Hobbes, *Leviathan* (London: Oxford University Press, 1909), I 13, p. 97.

23. *Metaphysics of Morals*, p. 124 (Ak. VI, pp. 312–13).

24. Kant writes: "It is not experience from which we learn of men's maxim of violence. . . . On the contrary . . . before a public lawful condition is established, individual men . . . can never be secure against violence from one another, since each has its own right to do *what seems right and good to [him]* and not to be dependent upon another's opinion about this" (ibid.)

25. "Theory and Practice," p. 70 (Ak. VIII, p. 289).

26. Ibid., p. 74 (Ak. VIII, p. 293). See also "What is Enlightenment?," in *Foundations of the Metaphysics of Morals*, transl. Lewis White Beck (New York: Macmillan, 1985), p. 85 (Ak.VIII, p. 35): "Laziness and cowardice are the reasons why so great a portion of mankind, after nature has long since discharged them from external direction (*naturaliter maiorennes*), nevertheless remains under lifelong tutelage." Right cannot be grounded in the difference between tacit and explicit consent, in the sense of Locke, because this would amount to admit the prior role of ready-made determination in reflection.

27. "Theory and Practice," p. 77 (Ak. VIII, p. 297).

28. Ibid.

29. Ibid., p. 84 (Ak. VIII, p. 306).

30. Ibid., p. 82 (Ak. VIII, p. 304).

31. Ibid., p. 76n. (Ak. VIII, p. 295n).

32. *Leviathan*, I 13, p. 95.

33. "Theory and Practice," p. 73 (Ak. VIII, p. 291).

34. Ibid., p. 79 (Ak. VIII, p. 299).

35. Ibid., p. 82 (Ak. VIII, p. 303).

36. In this connection, the reader will recall that Kant's project for perpetual peace remains perfectly compatible with the existence of violent, bloody battles. In this text, Kant claims that evil lies in the endless *preparations* for war, not in war as such.

37. Schelling, *System of Transcendental Idealism*, p. 221.

38. Ibid., p. 225.

39. Ibid., p. 223.

40. Schelling, *Further Presentations from the System of Philosophy* (1802), quoted in Heidegger, *Schelling's Treatise on the Essence of Human Freedom*, pp. 45–46.

41. "What is Enlightenment?," p. 86ff. (Ak. VIII, p. 36ff.)

42. Ibid., p. 88 (Ak. VIII, p. 39).

43. *The Conflict of the Faculties*, Ak. VII, p. 85.

44. F. Nietzsche, *The Gay Science*, sec. 344.

45. *Kant and the Problem of Metaphysics*, p. 156.

46. Ibid., p. 159.

Index

Q

R

S

T